The Influence of Global Ideas on Environmentalism and Human Rights

Markus Hadler

The Influence of Global Ideas on Environmentalism and Human Rights

World Society and the Individual

palgrave
macmillan

Markus Hadler
Department of Sociology
University of Graz
Graz, Austria

ISBN 978-1-137-57439-8 ISBN 978-1-137-57440-4 (eBook)
DOI 10.1057/978-1-137-57440-4

Library of Congress Control Number: 2017950866

Cover illustration: hope1983/Getty

Printed on acid-free paper

This Palgrave Macmillan imprint is published by Springer Nature
The registered company is Macmillan Publishers Ltd.
The registered company address is: The Campus, 4 Crinan Street, London, N1 9XW, United Kingdom

Preface and Acknowledgments

Human rights and environmentalism are global topics. Similarly, the development of this book has an inherent international dimension. It ties together topics that I have worked on in the USA, at Stanford University and Marshall University; in Australia, at Macquarie University; and in Austria, at the University of Graz.

At all these places, I have received valuable feedback from various scholars and friends. At Stanford University from members of the Comparative Workshop and its organizer, John W. Meyer, at Marshall University from Marty Laubach and Richard Garnett, at Macquarie University from Shaun Wilson and Jonathan Symons, and at the University of Graz from Max Haller, Franz Höllinger, and many others.

Work on this book and related research was also supported by different countries. My initial research visit to Stanford University and the Austrian 2010 environment survey were funded by the Austrian Science Fund, work on the final manuscript was funded by Macquarie University via their outside study program, and research and conference visits were supported by all of the aforementioned universities.

Finally, my involvement in international comparative research on environmentalism does not come to an end with this book. At the 2017 meeting of the International Social Survey Programme in Lausanne, we decided to repeat the survey on environmental attitudes and behaviors in

2020. I was elected to chair the development of this questionnaire, which guarantees heightened engagement with this topic.

Graz, Austria

Markus Hadler

Contents

About the Author

Markus Hadler is Professor of Sociology, at the University of Graz, Austria, and Honorary Professor at the Department of Sociology, Macquarie University, Australia. He is also the Austrian representative to the International Social Survey Programme (ISSP) and editor-in-chief of the *International Journal of Sociology*. His research interests lie in the areas of social inequality, political sociology, and environmental sociology.

List of Figures

List of Tables

1

Introduction

Scanning through time.com articles[1] with an eye on environment and climate, the year 2016 can be described by the following news articles: '2016 Was the Hottest Year Ever. It's the Third Year in a Row of Record-High Heat,' 'More Than 170 Countries Sign Paris Agreement in a Landmark For The Climate Fight,' 'Eight Eco-Conscious Instagram Accounts to Follow on Earth Day,' and 'These Kids Are Suing the Federal Government to Demand Climate Action.' These four reports show the whole spectrum of (a) the global problem, (b) international governmental action, (c) the global idea of 'Earth Day,' and (d) legal action as one possible means of action in civil society.

Searching through this archive with a focus on human rights would surface similar themes and articles. However, the goal of this book is not to search through this archive, but rather to ask if international efforts and global ideas result in any change. The sociological frame is 'World Society' theory, a sociological approach that commits itself to conceptualizing the effects of global ideas on organizational outcomes. It was developed by scholars working with John W. Meyer at Stanford University (see Kruecken and Drori 2009; Meyer 2009; Meyer et al. 1997).

M. Hadler, *The Influence of Global Ideas on Environmentalism and Human Rights*,
DOI 10.1057/978-1-137-57440-4_1

World Society theory—also known as World Polity (see Cole 2017)—has become one of the dominant approaches in international comparisons for studying the influence of global ideas on nation-states, organizations, and other actors around the world. It focuses on the diffusion of global ideas as cultural-cognitive scripts, which then constitute the behavior and decisions of actors. These considerations have been applied to various subjects, with the main areas of interest being human rights, environmentalism, and education.

Research in this vein has shown that world societal ideas—here used in the sense of global ideas and not as assumptions of this theory[2]—have shaped the adoption of human rights treaties, environmental protection outcomes, curricula content, and many other outcomes (Boyle et al. 2015; Bromley et al. 2011; Cole 2012b; Frank et al. 2000a; Hironaka 2014; Longhofer and Schofer 2010; Meyer et al. 1997). Such an influence of international organizations and treaties is obvious when considering coercion in terms of laws and international binding treaties. A second diffusion mechanism, or pillar using Scott's (2014) terminology, would be normative aspects. Here, we can think of standardization processes such as International Organization for Standardization (ISO) norms that offer organizational templates for different aspects of organizations and social life, without being legally enforced.

The third pillar is cultural ideas and models, which are in the center of World Society theory (Scott 2014). Here, we can think of the effects of global ideas such as the aforementioned Earth Day. Such ideas are not enforced by any governmental agency nor do they aim to standardize life. Their effects on organizations and social life are therefore less obvious. This was reflected in the debate between Frank et al. (2000b) and Brulle (2000) on the question of whether the international level influences environmental outcomes. Frank et al.'s (2000b) conclusion was that that some parts work, while others do not, but that it is better to have a pockmarked system than no system at all. Since then, World Society scholars have continued to look into the effects of the international level and more evidence for its effects has surfaced (see Hironaka 2014).

Support for the idea that the international level exerts effects also comes from outside academia. Considering the recent moves in Russia and other countries to force non-governmental organizations (NGOs) who receive funding from abroad to register as international agents clearly shows that governments worry about the influence of such groups. These moves thus corroborate scholarly discussions about effects such as the boomerang effect (Keck and Sikkink 1998) where local activists pressure the government from below, but at the same time, also motivate the international level to become active, which results in twice the pressure on governments.

Much of the World Society related research focuses on institutional outcomes such as the funding of environmental ministries and parks or human rights institutions such as an Ombudsman office. Scholars ask less frequently if international efforts and global ideas also affect individual attitudes and behaviors. This shortcoming is surprising, given that World Society evolved from micro-sociological accounts of Schütz, Luckmann, and others by applying their ideas on organizations (see Meyer 2007; Scott 2014).

The few studies that link individual attitudes and behaviors to global ideas often resort to a simple exposure model, with the idea that 'the stronger the ties of a country to the global community, the stronger the influence on an individual' (Givens and Jorgenson 2013; Knight and Messer 2012; Marquart-Pyatt 2012). This simple exposure model, however, has been extended in research on organizational change, where phenomena such as isomorphism, diffusion, and decoupling have been discussed in detail (Bromley and Powell 2012; Meyer 2007).

Applying the simple exposure idea and considering the multiple efforts of the international community to raise the global awareness of environmental and human rights issues, one would thus expect that environmental concerns and attitudes toward human rights issues would increase globally. However, when looking at public opinion data on environmental actions—see Chap. 6—we see that public environmental behavior has declined in most countries or remained at a low level. In human-rights-related attitudes such as tolerance toward people of a particular race or sexual orientation—see Chap. 7—we see improvements in some countries, but also backlashes in other countries.

These mixed trends in the empirical findings suggest that the connection between the global level and individual attitudes is more complex than a straightforward exposure effect. It also questions the dominant narrative of World Society related research of a positive influence of global ideas on various outcomes. This book aims to address this conundrum by conceptualizing and empirically testing the links between global ideas and individual outcomes. It focuses on two core topics of World Society: human rights and environmentalism, with greater emphasis on environmentalism.

Environmentalism and human rights issues have more in common than it might seem at the first glance. Environmentalism has roots that go back to the nineteenth century, but its takeoff occurred only in the 1960s and 1970s when pollution became more obvious and value shifts toward perceiving the environment as an entity of its own became important (Harper 2012; Rootes and Brulle 2013). Human rights also has a long history, but did not gain wide recognition internationally before the Universal Declaration of Human Rights after World War II and the related international covenants that came into being in the following decades (Kennedy 2006).

Furthermore, environmental concerns started out with rather narrow conservation and preservation topics and later shifted to more encompassing deep ecology and environmental justice issues (Harper 2012). Human rights, as Brysk (2013) summarizes, started with a focus on civil and political rights but extended in the 1980s to issues of economic and social rights, including problems due to economic exploitation and development. Most recently, their scope also extended to topics such as sexual orientation. Environmentalism and human rights also share a global dimension, which transcends the nation-state.

Environmental problems such as nuclear fallout and acid rain cannot be stopped by national borders and in addition they threat all groups within a given society (Beck 1986). Human rights are assigned to all individuals, are not based on citizenship of a specific country, and can be claimed at international bodies (Kennedy 2006). Finally, human rights and environmental topics have started to overlap. The frame of environmental justice refers specifically to an uneven share of environmental burdens within and between societies, and NGOs have started to take up topics from both areas or cooperate, as can be seen in the collaboration of Sierra Club with Amnesty International in an effort to protect environmental activists around the globe (Brysk 2013).

1.1 Alternative Views

The emphasis of this book on the international level must not distract from the fact that other forces also shape individual attitudes and behaviors in the areas of environmentalism and human rights. Among these, the two most prominent views are modernization theory and social movement theory, which are frequently combined with World Society (see Boyle et al. 2002; Hironaka 2014; Smith and Wiest 2012). This section now summarizes the main ideas of these two approaches, whereas the more specific strands are addressed in the empirical chapters of this book.

A difference between World Society and modernization theories is that the latter concentrate on endogenous developmental processes of societies. Changes in individual beliefs and attitudes are considered independent from the developments in other nations and the relationship to other societies (Boyle et al. 2002). Modernization processes themselves include various aspects, for example economic development and democratization (Almond and Verba 1965; Lipset 1994).

The modernization theories of the post-war era emphasized the macro level; the impact on individuals was only of secondary importance. Dominated by the theory of Talcott Parsons (1972), it was assumed that all countries follow similar development processes, with the USA as a role model. Parsons emphasized the structural–functional differentiation of societies, which divides societies into different subsystems such as the economic, the political, or the social system. This subdivision of societies into different spheres is extremely useful for comparing societies (see e.g. Gerhards and Hoelscher 2003; Muench 1986), and it guides the country descriptions in Chap. 5.

The functionalist theory, however, has an emphasis on the societal level and minor interest in the impact on the individual. Various authors have attempted to overcome this shortcoming. Ulrich Beck (1986, 1993), for example, asserts that a 'second modernity' has emerged and emphasizes its non-intended impacts on individuals. He criticizes the assumptions of simple modernization theory, such as linear progress, instrumental rationality, or functional differentiation. Beck also asserts that risks have become global; therefore national borders have become obsolete—an aspect that is of particular importance for environmentalism, as pointed out above.

Another well-known approach is Ronald Inglehart's (1989) work on post-materialism. Inglehart asserts that economic development results in a shift of individual beliefs and attitudes toward more openness and liberal values. 'Industrialization promotes a shift from traditional to secular-rational values, while the rise of post-industrial society brings a shift toward more trust, tolerance, well-being, and post-materialist values' (Inglehart and Baker 2000, 49). The core of these attitudes is the materialism—post-materialism dimension. Inglehart (1989) emphasizes that this dimension is correlated with a broad range of other attitudes in work, politics, self-expression, and—in line with the themes of this book—increasing environmentalism and openness to diversity.

Inglehart's theory (1989) is applied to both the societal and the individual level, with affluent societies characterized by a larger percentage of post-materialists, and individuals who are well-off are more likely to be post-materialists. Societal and individual traits can also interact. Inglehart has also shown that younger cohorts are more often post-materialists in both affluent and less affluent societies, but that the general strength of post-materialistic orientations is higher within rich societies.

While Inglehart's theory is widely used, it is also contested (see e.g. Haller 2002). As for environmentalism, a dispute rages over the influence of environmental pollution (objective dimension) versus that of subjective values (see Brechin and Kempton 1997; Dunlap and Mertig 1997; Kidd and Lee 1997). Subsequently, Brechin (1999) analyzed the roles of post-materialism and the actual state of the environment as the cause of environmental concerns in developed and developing countries. His findings suggest that worries about local problems are more widespread in southern countries while worries about the global environmental state are more or less similar in all countries. On the other hand, Oreg and Katz-Gerro's (2006) results suggest that the national level of post-materialism positively affects environmental concern.

Equally important as modernization approaches are approaches that discuss political action within the framework of social movements. This literature focuses on factors that promote and hinder social movements and related behaviors. It considers three distinct aspects: political opportunities, resources, and framing. Social movements are more common in open democracies and, within democracies, dependent on the specific political circumstances. (McAdam et al. 1996, 2001; Snow et al. 2004; Tilly 2004; van der Hiejden 2006). In line with this view, as Chap. 5 will discuss, environmental movements flourished in Western countries from the 1970s onward, whereas their takeoff in the socialist systems of Eastern Europe occurred only during the transition period from socialism to capitalism.

The second crucial factor in the success of social movements is resource mobilization and the social appropriation of means (Edwards and McCarthy 2004; McCarthy and Zald 1977; McAdam et al. 1996, 2001; Tilly 2004). Resources cover various repertoires of action and available means; two important factors at the national level are monetary and cultural means. In affluent societies, social movements can draw upon ample financial resources, which make it easier to start and sustain movements (Gillham 2008). Among cultural means, education plays a major role since knowledge of issues is an important factor in environmental behavior (Dunlap and Mertig 1997).

Location is also important. International movements are often located in the same place as their opponents (Tilly 2004) and, thus, can be found more often in cities such as Brussels, London, or New York. Urban areas, in general, are also attractive since social movements are also dependent on human resources and it is likely to find activists in such places (Gillham 2008). Protests thus happen more frequently in cities, with the exception of direct action at sites of environmental threats such as nuclear power plants, dams, and similar locations.

Thirdly, the success of social movements also depends on the framing of an issue (Benford and Snow 2000; McAdam et al. 2001). Framing refers to the cognitive and emotional dimension of social action; actors are motivated to take action only if a need for action is perceived. As far as environmental movements are concerned, environmentally damaging incidents, such as chemical spills in the Rhine, acid rain, the nuclear reactor incident in Chernobyl, kept them flourishing (Gillham 2008) or as in the case of grassroots movements, such incidents initiated the actual movement. Chapter 5 will include a brief overview of perceived environmental threats and the state of the environment in selected countries.

1.2 Content of This Book

The alternative views mentioned above inform this book in various places. Its main structure, however, follows the goal of conceptualizing World Society theory, the individual actor, and the influence of global ideas on environmentalism and attitudes toward human rights. For this purpose, the first half of this book discusses the theoretical underpinning of World Society and the structure of the international level, whereas the second half presents empirical results on the effects of ties to world society.

Chapter 2, 'World Society and Neo-Institutionalism,' places World Society theory within the larger picture of neo-institutionalism. It discusses its main assumptions, the rise and use of this theory, and its relation to other neo-institutional approaches in economics and political science. It shows that World Society's distinctive characterization is a focus on cultural-cognitive elements, which were initially taken from micro-sociological accounts and applied to organizational structures and change.

The roots in micro-sociological accounts resulted in a terminology that is filled with words such as the actor, the agent, the script followers, and so on. The empirical focus of World Society theory, however, has shifted to an emphasis on the macro level and resulted in a neglect of the individual. Chapter 3, 'World Society and the Individual,' thus discusses the remaining elements influencing individual actors and describes how global ideas can influence attitudes and actions.

The chapter also includes a summary of existing studies on the link between global ideas and individual attitudes and actions. This overview shows that these studies often rely on a simple exposure model, with age, place of residency, and education considered as important mediators and moderators of world societal models. Yet conflicts between the prevalent views about the influence of the immediate surroundings of the individual and coercion via regulative institutions are possible and have been addressed in recent literature.

Chapter 4, 'The International Level: IGOs and INGOs,' summarizes the main developments in the areas of human rights and environmentalism, with an emphasis on the United Nations, related treaties, and NGOs such as Greenpeace, Amnesty International, and Human Rights Watch.

This chapter addresses the critique that World Society related research often reduces the impact of global ideas and institutions to the indicators of numbers of International Governmental Organizations (IGOs) and International NGOs (INGOs) without acknowledging the developments in this sphere.

The chapter on the international level shows that environmentalism goes back to the late nineteenth century when various conservation and preservation movements became active. The rise of more contemporary movements, however, occurred only in the 1960s and 1970s, accelerated by the dire predictions of the Club of Rome, visible environmental pollution, value change, and other factors. As for human rights, the initiation at the international level occurred with the Universal Declaration of Human Rights after World War II. Ever since, the scope of human rights has been extended and now encompasses discrimination based on sexual orientation, age, or health.

After discussing developments at the international level, Chap. 5, 'Development in Selected Countries,' discusses the specific circumstances in Austria, Canada, Czech Republic, Germany, Russia, and the USA with a focus on the environment. For each country, relevant political, economic, and social developments are discussed and related to changes in environmental attitudes and behaviors. The conclusion of this chapter is that the specific political and economic circumstances do have an influence on environmental attitudes, yet all countries are characterized by similar trends over time, which point to the relevance of the global level.

Chapter 6, 'Individual Environmentalism,' presents the results of empirical analyses on the influence of global ideas and related factors on individual environmental behaviors. The analyses are based on public opinion data collected by the International Social Survey Programme in 1993, 2000, and 2010. The overall trends show declining support for environmental organizations and increasing private pro-environmental behaviors. Multi-level models are afterward used to simultaneously estimate the influence of ties to world society, national characteristics, and individual characteristics on these environmental concerns. It shows that global ideas have a stronger influence on public behavior than that on the private dimension of environmental behaviors.

Chapter 7, 'Individual Attitudes Toward Human Rights,' looks into xenophobia and homophobia across countries and time. The empirical analyses use data collected by the World Values Survey group (1981–2014) and Pew Research Center (2002–2013). The descriptive analyses show divergent trends depending on the subject of human rights—for example increasing tolerance toward LGBT (lesbian, gay, bisexual, and transgender) rights and groups, stagnant tolerance toward immigrants, and persistent differences between geo-political regions. As in Chap. 6, multi-level models are used to estimate the influence of world societal ideas, national characteristics, and individual characteristics on attitudes toward different human rights topics. These models show that global ties have a stronger influence on homophobia than on xenophobia.

The final chapter (Chap. 8, 'Conclusion') summarizes the findings across all themes and topics of this book. It names examples for the influence of regulative, normative, and cultural-cognitive elements of institutions on environmentalism and attitudes toward human rights. It casts some doubt on the overall applicability of the exposure model and concludes that the ties to the global level have stronger effects on topics that gained salience more recently. This, however, does not exclude the possibility that these forces had a powerful role in the past in establishing now taken-for-granted political and civic rights.

Notes

1. http://time.com/4637064/2016-hottest-year-on-record/, http://time.com/4304724/paris-agreement-climate-change-signing/, http://time.com/4300678/earth-day-instagram/?iid=sr-link3, http://time.com/4567012/federal-government-lawsuit-climate-change/?iid=sr-link5 (accessed 20 Apr 2017).
2. World society can refer to the sociological theory 'World Society,' but also to the coalescence of cultural models, ideas, and organizations at the global level, whose influence on organizations and actors is conceptualized within this theory. The capitalized term 'World Society' is used when referring to the theory, whereas world society refers to global ideas and models.

References and Further Readings

Almond, G.A., and S. Verba. 1965. *The Civic Culture: Political Attitudes and Democracy in Five Nations*. Boston, MA: Little, Brown.

Beck, U. 1986. *Risikogesellschaft. Auf dem Weg in eine andere Moderne*. Frankfurt am Main: Suhrkamp.

———. 1993. *Die Erfindung des Politischen*. Frankfurt am Main: Suhrkamp.

Benford, R.D., and D.A. Snow. 2000. Framing Processes and Social Movements: An Overview and Assessment. *Annual Review of Sociology* 26 (1): 611–639.

Boyle, E.H., B.J. McMorris, and M. Gomez. 2002. Local Conformity to International Norms: The Case of Female Genital Cutting. *International Sociology* 17 (1): 5–33.

Boyle, E.H., K. Minzee, and W. Longhofer. 2015. Abortion Liberalization in World Society, 1960–2009. *American Journal of Sociology* 121 (3): 882–913.

Brechin, S.R. 1999. Objective Problems, Subjective Values, and Global Environmentalism: Evaluating the Postmaterialist Argument and Challenging a New Explanation. *Social Science Quarterly* 80 (4): 793–808.

Brechin, S., and W. Kempton. 1997. Beyond Postmaterialist Values: National Versus Individual Explanations of Global Environmentalism. *Social Science Quarterly* 78 (1): 16–20.

Bromley, P., J.W. Meyer, and F.O. Ramirez. 2011. The Worldwide Spread of Environmental Discourse in History, Civics, and Social Studies Textbooks. *Comparative Education Review* 55: 517–545.

Bromley, P., and W. Powell. 2012. From Smoke and Mirrors to Walking the Talk: Decoupling in the Contemporary World. *The Academy of Management Annals* 6: 483–530.

Brulle, R.J. 2000. *Agency, Democracy, and Nature. The US Environmental Movement from a Critical Theory Perspective*. Cambridge: MIT Press.

Brysk, A. 2013. Human Rights Movements. In *The Wiley-Blackwell Encyclopedia of Social and Political Movements* (Online).

Cole, W.M. 2017. World Polity or World Society? Delineating the Statist and Societal Dimension of the Global Institutional System. *International Sociology* 32 (1): 86–104.

Dunlap, R.E., and A.G. Mertig. 1997. Global Environmental Concern: An Anomaly for Postmaterialism. *Social Science Quarterly* 78 (1): 24–29.

Edwards, B., and J.D. McCarthy. 2004. Resources and Social Movement Mobilization. In *The Blackwell Companion to Social Movements*, ed. D.A. Snow, S.A. Soule, and H. Kriesi, 116–152. Oxford: Blackwell.

Frank, D.J., A. Hironaka, and E. Schofer. 2000a. The Nation-State and the Natural Environment over the Twentieth Century. *American Sociological Review* 65 (1): 96–116.

———. 2000b. Environmentalism as a Global Institution: Reply to Buttel. *American Sociological Review* 65 (1): 122–127.

Gerhards, J., and M. Hoelscher. 2003. Kulturelle Unterschiede zwischen Mitglieds- und Beitrittsländern der EU. Das Beispiel Familien- und Gleichberechtigungsvorstellungen. *Zeitschrift für Soziologie* 32: 206–225.

Gillham, P. 2008. Participation in the Environmental Movement: Analysis of the European Union. *International Sociology* 23: 67–93.

Givens, J., and A.K. Jorgenson. 2013. Individual Environmental Concern in the World Polity: A Multilevel Analysis. *Social Science Research* 42: 418–431.

Haller, M. 2002. Theory and Method in the Comparative Study of Values. Critique and Alterative to Inglehart. *European Sociological Review* 18: 139–158.

Harper, Charles. 2012. *Environment and Society. Human Perspectives on Environmental Issues*. 5th ed. Boston, MA: Pearson.

Hironaka, A. 2014. *Greening the Globe. World Society and Environmental Change*. Cambridge: Cambridge University Press.

Inglehart, R. 1989. *Kultureller Umbruch. Wertwandel in der westlichen Welt*. New York: Campus Verlag.

Inglehart, R., and W.E. Baker. 2000. Modernization, Cultural Change, and the Persistence of Traditional Values. *American Sociological Review* 65: 19–51.

Keck, M., and K. Sikkink. 1998. *Activists Beyond Borders*. Ithaca, NY: Cornell University Press.

Kennedy, P. 2006. *The Parliament of Man. The United Nations and the Quest for World Government*. London: Allen Lane.

Kidd, Q., and A. Lee. 1997. Postmaterialist Values and the Environment. A Critique and Reappraisal. *Social Science Quarterly* 78 (1): 1–15.

Knight, K., and B. Messer. 2012. Environmental Concern in Cross-National Perspective: The Effects of Affluence, Environmental Degradation, and World Society. *Social Science Quarterly* 93: 521–537.

Kruecken, G., and G. Drori, eds. 2009. *World Society: The Writings of John W. Meyer*. Oxford: Oxford University Press.

Lipset, S.M. 1994 [Original 1959 and 1981]. *The Political Man. The Social Bases of Politics*. Baltimore, MD: The John Hopkins University Press.

Longhofer, W., and E. Schofer. 2010. National and Global Origins of Environmental Association. *American Sociological Review* 75: 505–533.

Marquart-Pyatt, S.T. 2012. Contextual Influence on Environmental Concern Cross-Nationally: A Multilevel Investigation. *Social Science Research* 41: 1085–1099.

McAdam, D., J. McCarthy, and M.N. Zald, eds. 1996. *Comparative Perspectives on Social Movements: Political Opportunities, Mobilizing Structures, and Cultural Framings*. New York: Cambridge University Press.

McAdam, D., S. Tarrow, and C. Tilly. 2001. *Dynamics of Contention*. Cambridge: Cambridge University Press.

McCarthy, J.D., and M.N. Zald. 1977. Resource Mobilization and Social Movements: A Partial Theory. *American Journal of Sociology* 82: 1212–1241.

Meyer, J.W. 2007. Reflections on Institutional Theories of Organizations. In *Handbook of Organizational Institutionalism*, ed. R. Greenwood, C. Oliver, R. Suddaby, and K. Shalin-Andersson, 788–809. Thousand Oaks, CA: Sage.

———. 2009. Reflections: Institutional Theory and World Society. In *World Society: The Writings of John W. Meyer*, ed. G. Krucken and G.S. Drori, 36–63. Oxford: Oxford University Press.

Meyer, J.W., J. Boli, G.M. Thomas, and F.O. Ramirez. 1997. World Society and the Nation-State. *American Journal of Sociology* 103: 144–181.

Muench, R. 1986. *Die Kultur der Moderne. Ihre Grundlagen und ihre Entwicklung in England und Amerika*. Frankfurt am Main: Suhrkamp.

Oreg, S., and T. Katz-Gerro. 2006. Predicting Proenvironmental Behavior Cross-Nationally. Values, the Theory of Planned Behavior, and Value-Belief-Norm Theory. *Environmental Behavior* 38: 462–483.

Parsons, T. 1972. *Das System moderner Gesellschaften*. Weinheim: Juventa.

Rootes, C., and R. Brulle. 2013. Environmental Movements. In *The Wiley-Blackwell Encyclopedia of Social and Political Movements* (Online).

Scott, W.R. 2014. *Institutions and Organizations: Ideas. Interests, and Identities*. 4th ed. Thousand Oaks, CA: Sage.

Smith, J., and D. Wiest. 2012. *Social Movements in the World System. The Politics of Crisis and Transformation*. New York: Russel Sage Foundation.

Snow, D.A., S.A. Soule, and H. Kriesi, eds. 2004. *The Blackwell Campanion to Social Movements*. Malden, MA: Blackwell.

Tilly, C. 2004. *Social Movements, 1768–2004*. Boulder, CO: Paradigm Publishers.

Van der Hiejden, H.-A. 2006. Globalization, Environmental Movements, and International Political Opportunity Structures. *Organization and Environment* 19: 28–45.

2

World Society and Neo-institutionalism

This chapter offers an introduction to World Society theory and its main assumptions and ideas. It starts with a basic discussion of the term 'institution,' which is used in many different ways and can refer to basic human actions or to types of organizations. Next is an overview of the historical developments in various social sciences that have led to the development of neo-institutional theories.[1] In this regard, World Society theory is part of the neo-institutionalist turn of the 1960s and 1970s, which has affected sociology, political science, economics, and other disciplines. This chapter summarizes the developments in these three disciplines by pointing out the differences in their approaches. After looking at the developments across social sciences, the focus turns to World Society theory. We identify its roots in the early papers of Meyer and Rowan, the main areas of its application, the dominant empirical approaches, and common misperceptions about this theory. The conceptualization of the individual is briefly mentioned here, but considered in depth in Chap. 3.

M. Hadler, *The Influence of Global Ideas on Environmentalism and Human Rights*,
DOI 10.1057/978-1-137-57440-4_2

2.1 Institutions

The different institutionalisms across social sciences are characterized by a common interest in the effects and origins of 'institutions.' The term 'institution' can be used in various regards and, according to Immergut (2011), refers to the degree of formalization of social behavior, ranging from casual habits to formal arrangements, to the extent of the collective binding, ranging from traditions and folkways to sanctioned laws, and to the way they are sanctioned, from self-reinforcement to coercion by other entities. Finally, 'institution' can also refer to organizations, such as institutes of scientific studies, or organizations dealing with 'institutions' such as education and monetary instruments.

In terms of scientific disciplines, we can distinguish between early institutionalism and neo-institutionalism. Early institutionalism refers to approaches in the late nineteenth and early twentieth centuries and scholars such as Thorsten Veblen and Max Weber. Their main emphasis was on studying the development of large formal institutions such as legal, economic, religious, and other societal systems. These approaches lost ground in the early decades of the twentieth century, when behavioristic models gained importance. Once shortcomings and inconsistencies in these behavioristic models became obvious, institutionalism was revived in various disciplines, starting in the 1970s.

These new institutionalisms share only some of the views of the early institutionalism, despite being labeled 'neo'-institutionalism. A main difference is that neo-institutionalism considers institutions broader than the earlier scholars and rather used Peter Berger and Thomas Luckmann's definition that 'institutionalization occurs whenever there is a reciprocal typification of habitualized action by types of actors' (1966, 54). In this regard, the term 'institution' is not limited to formal institutions any more, but emphasizes the experience of an institutional world as an objective reality. There are, however, significant differences within the contemporary 'neo-institutionalisms,' which will be addressed in the following sections (see also Greenwood et al. 2008; Powell and DiMaggio 1991; Schofer et al. 2012; Scott 2014).

2.2 The Developments in Different Social Sciences

Economics

The early institutionalism in economics developed in opposition to the neo-classical view that individuals are rational economic actors and that a few basic assumptions and laws are sufficient to explain economic processes. These conflicting views found their expression in the German Methodenstreit of the late nineteenth century, in which institutionalists such as Gustav Schmoller emphasized that economic processes are outcomes of specific historical developments and embedded in cultural settings, whereas scholars such as Carl Menger represented the neo-classical views. Early American institutionalists such as Thorstein Veblin and John Commons were influenced by this German discussion and shared the interest in the historical development of economic phenomena.

Thorstein Veblen in his book *The Theory of the Leisure Class: An Economic Study of Institutions* (1899) studied individual over-consumerism, a behavior that violates the assumption of rational, resourceful, and efficient actions. Veblen identified the stratification system and social classes in feudalism as a possible cause of this utility violating behavior. John Commons, in contrast, focused on the transaction, the basic social interaction of two parties. This interaction follows certain rules of conduct, which are also considered social institutions. Commons was particularly interested in the historical development and origin of these institutions and summarized his views on the collective nature of institutions in his 1934 book *Institutional Economics.*

The early institutionalism lost ground in economics over the first decades of the twentieth century, while neo-classic views gained influence. A possible explanation for the demise of early institutionalism can be seen in the institutionalists' overemphasis on historical developments and idiosyncratic explanations with little emphasis on theory development. This diminished influence is also reflected in the fact that Ronald Coase's 1937 article, 'The Nature of the Firm,' that focused on transaction costs and is now considered a classic work by neo-institutionalists, went rather unrecognized for several decades.

It was only in the 1970s, when Oliver Williamson added specific conditions such as increasing transaction costs under uncertainty and individual opportunism to Coase's work, that this work became more popular. It was also Williamson (1975) who coined the term 'economic neo-institutionalism.' Williamson's work, however, shows an important difference to the early institutionalism and highlights its closeness to neo-classic views: Williams assumes that individuals have given preferences and does not consider individuals as being shaped by specific historical developments and a cultural embedding.

This difference is also generally reflected in contemporary economic neo-institutionalism, which shares only some assumptions with early institutionalism. Similar to the early approaches, it focuses on the effects of institutions and their origin and evolution. It considers 'Institutions … written and unwritten rules, norms, and constraints that humans devise to reduce uncertainty and control their environment' (Menard and Shirley 2005, 1). However, it also adopted the neo-classic assumptions of scarcity and competition, without the notion of perfect information and instrumental rationality.

A different strand within economics focuses on the macro development of societal institutions and thus is close to historical institutionalism, which will be discussed in the political science section. A prime example in this regard is the economic historian Douglas North, who, for example, discussed the types and forms of economic exchange in different societies and how their patterns and structures have evolved over time (North 1991).

The historical development of contemporary economic institutions is also addressed in evolutionary economics. Richard Nelson and Sidney G. Winter's book *An Evolutionary Theory of Economic Change* (1982) that describes changes in routines and technology is an example of this. Despite not explicitly referring to institutions, their research resonates with the ideas of Thorstein Veblen and particularly Joseph Schumpeter, one of the few economists who continued to work in the tradition of early institutionalism.

Political Science

Institutional approaches with a focus on descriptions of the roots and development of legal systems dominated in political science in the nineteenth and early twentieth century. Prominent scholars in this time were J. W. Burgess and the later US President Woodrow Wilson. These scholars produced dense descriptions of legal systems and other political institutions that were guided by moral philosophical accounts and—similar to the early institutionalism in economics—rather scarce in theory building. This approach lost ground in the first decades of the twentieth century when behavioral perspectives gained importance, accompanied by a topical shift to political behavior, voting behavior, and public opinion.

Ideas of these early institutionalists were revived in a reaction to group theories and structural-functionalist theories that were dominant in the 1960s and 1970s and labeled historical institutionalism. Historical institutionalism sees institutions as '… the formal or informal procedures, routines, norms and conventions embedded in the organizational structure of the polity or political economy' (Hall and Taylor 1996, 938).

Among others, the rise of this theory in political science is associated with Theda Skocpol's book *States and Social Revolutions: A Comparative Analysis of France, Russia and China* (1979) which focuses on the social institution 'state' and its influence on social change. The interest in the development of modern institutions resonates with the ideas of early institutionalists, but differs as it does not follow normative considerations. Outcomes are not considered predetermined, but often as unintended consequences. Insofar, there is no necessary trend toward democracy, modernity, and other societal traits. Institutions, however, are not considered the only determinant of social change. Factors such as dominant ideas and societal affluence are taken into account as well.

Despite being a reaction to group theories and structural-functionalism, the historical institutionalism nonetheless incorporated ideas of these theories (Hall and Taylor 1996). It kept the assumption of rivalry and scarcity. In this regard, power differences between groups are a central concern, and historical institutionalists frequently focus on the role of institutions in giving some groups more power than others. Similar to structural-functionalists, historical institutionalists consider the polity a system of interrelated parts, but rather focus on its influence on individual behaviors rather than taking individual traits and beliefs as the basis of this system.

As far as the relationships between individuals and institutions are concerned, historical institutionalism is located between rational-choice and sociological neo-institutionalism (Hall and Taylor 1996). Scholars use both a rational-choice-related approach, where institutions provide information that can reduce the uncertainty for actors who make rational choices, and a cultural approach, where institutions are considered cognitive templates that frame the choices of an individual. While the former considers individuals utility maximizers, the latter sees them as individuals who select or satisfy different options.

Alongside this historical institutionalism, a rational-choice-focused neo-institutionalism that extends economic institutionalism to political science has developed as well. The rise of this rational-choice neo-institutionalism was inspired by the paradox observation that the voting outcomes and agreements within the US Congress were relatively stable, whereas rational choice theory would predict constantly changing alliances, given the many different interests of legislators (Hall and Taylor 1996). An explanation of these stable outcomes is 'institutions' as they are considered to 'lower the transaction costs of making deals' (ibid, 943). These ideas have been employed well beyond the US Congress and applied to the influence of laws, political institutions, and their evolution.

Sociology

Institutions have been discussed within sociology in more detail than in economics and political science and, thus, the perspectives of various classic European and American sociologists are reflected in contemporary neo-institutionalism (see Scott 2014).

On the European side, Karl Marx's view that the material world and the associated beliefs and norms determinate the social construction of reality, Emile Durkheim's emphasis on social facts that govern human behavior, and Max Weber's approach that considered economic institutions specific historical developments, but was also in favor of theory building, can be named. Among the American sociologists, Herbert Spencer's and William Sumner's division of societies into different segments, Everett Hughes' studies of the institutional structures of the workplace, Talcott Parsons' system of action, and Peter Berger and Thomas Luckmann's theory of the social construction of reality can be mentioned.

A first important shift toward the contemporary sociological neo-institutionalism was the application of these classic ideas to organizations—an aspect the early institutionalists with their focus on broader social phenomena such as political and economic systems did not consider explicitly. The impetus to focus on organizations came via Robert K. Merton and, particularly, his student Philip Selznick. Selznick's (1949) study of the Tennessee Valley Authority (TVA) can be considered a classic work in this regard. The TVA was founded by the US government to aid the underdeveloped areas between Kentucky and Alabama by acquiring private land to control floods and by producing electric power. The organization co-opted local actors who ultimately undermined its mission. Important features of this study are a focus on unintended consequences, political influences, and struggles, including those from outside the organization, and not just a narrow view on the economic environment.

In the consecutive decades, however, organizational sociology was dominated by contingence theory, resource dependency, and population ecology with a narrower focus on the economic environment (Senge 2007). Contingency theory, for example, focuses on technical and economic circumstances and which types of organizations survive; resource dependency on organizations that exchange resources and, even more restricted, on resources that have monetary value; and population ecology on the adaption of organizations. In the late 1970s and early 1980s, the interest in institutions gained importance. This new institutionalism revived perspectives of the classic approaches and added ideas developed in philosophy and social psychology.

The connection between neo-institutionalism and philosophy comes from phenomenology, a strand of philosophy that goes back to Edmund Husserl, a German scholar of the late nineteenth century. His ideas were picked up by various scholars and led, for example, to the development of philosophical anthropology, an approach that considers humans as both creatures and creators of their social environments and institutions a replacement for biological determinants. Phenomenology found its way into sociology via the Viennese scholar Alfred Schütz, whose social phenomenology was based on everyday knowledge. Sociologists such as Harold Garfinkel, Peter Berger, and Thomas Luckmann continued these thoughts. This focus on the semiotic functions of culture, with an emphasis on knowledge and belief systems, turned out to be central to sociological neo-institutionalism.

Parallel, neo-institutionalism also integrated developments of social psychology that emphasize the cognitive processes of individuals. Individuals are considered information processers that use particular 'scripts' and 'schemas' as templates in this interpretation. This view complements the sociological emphasis on the context that sees individuals as rather passively exposed individuals. It is also reflected in the organizational sociology definition of institutions: 'Institutions are socially constructed, routine-reproduced (ceteris paribus), program or rule systems. They operate as relative fixtures of constraining environments and are accompanied by taken for granted accounts' (Jepperson 1991, 149).

In sum, the old and the new institutionalism in sociology share various perspectives, but have also clear differences (see DiMaggio and Powell 1991, 12ff). Both are distant to rational actor models of organizations, emphasize the influence of the organizational environment, and consider the role of culture in shaping organizations. As for differences, early institutionalism focused on political influence and the resulting constraints, whereas the new institutionalism emphasizes the influence of common understandings that are taken for granted. Furthermore, the old institutionalism focuses on informal interactions within organizations to explain irrational decisions, whereas the new institutionalism considers the organizational structure itself irrational and decisions as driven by cultural accounts from outside the organization. It has also a broader understanding of organizational environments and does not limit itself to the local context. This is reflected in the new view that institutionalization occurs outside the organization, within industrial sections or even at the societal level. The classic organizational institutionalism, in contrast, located these processes within the organizations.

Differences Between the Three Main Areas

An obvious difference between these approaches first arises from the nature and topical areas of their disciplines. Less obvious differences also emerge regarding the role of the individual (Schofer et al. 2012). The economic rationalism operates under a rational actor assumption and asks why individuals enter institutional arrangements and studies the consequences of these arrangements on behaviors. Historical institutionalism focuses on macro developments and considers the evolution of different institutions, but places less emphasis on their influence on the individual (Hall and Taylor 1996). Sociological neo-institutionalism sees individuals as constituted by institutions, thus asserting that even the choices and ideas individuals perceive are shaped by the prevalent institutions (Jepperson 1991).

Another difference comes from the question of the origins of institutions. Both the sociological and the historical institutionalism assume an existing world of institutions. The question then arises how new institutions are drawn from this existing repertoire. As Hall and Taylor (1996) point out, the sociological view places less emphasis on the role of agents in implementing institutions, while the historical institutionalism places more emphasis on existing power relationships that favor some actors over others in the process of creating new institutions. Rational choice institutionalism, on the other hand, emphasizes the functions of institutions, and the reason why they were kept up.

Discursive Institutionalism

Recently, a fourth strand of neo-institutionalism has developed and is labeled discursive institutionalism or constructivist institutionalism (Schmidt 2008). This approach focuses on the role of ideas and is considered a more dynamic approach than the three other neo-institutionalisms. Rational-choice, historical, and sociological neo-institutionalism have a tendency to emphasize the external role of institutions in shaping individual behaviors, organizational goals, and other outcomes. The discursive approach, by contrast, considers institutions a process in which agents create and maintain institutions. In this regard, discursive institutionalism studies the development of ideas, places them in a specific context, and scrutinizes the way they are conveyed. In its application in political sciences, it is applied to two basic forms of discourse: between politicians (coordinate discourse) and between politicians and the general public (communicate discourse). In studying these processes, the research in the area of discursive institutionalism often integrates insights from the other three neo-institutionalisms and thus could be also considered a complementary approach.

2.3 World Society Theory

World Society has evolved from the early accounts of Meyer and Rowan (1977) on the connections between organizations and their environment, to a research program that conceptualizes and empirically tests the global diffusion of cultural models. World Society brings together three components: Firstly, in close relationship to other research on organizations, it builds upon the notion of organizations being constituted by ideas in their environment. Secondly, World Society emphasizes a specific set of cultural views—the Western cultural account—that has become influential over time and led to the dominance of specific global models that emphasize science, rationalization, actorhood, and so on. Thirdly, in combining these two aspects, it empirically researches the diffusion of such ideas and how they influence organizations and actors around the world.

World Society extends ideas developed within the new sociological institutionalism beyond the narrow scope of economic organizations by discussing the impact of global ideas on a variety of actors. Its core view is reflected in the hypothetical example Meyer et al. (1997) discuss in their article on world society and the nation-state: If explorers discovered a new island inhabited by a traditional society, one can easily imagine that it would immediately be visited by consultants of various international organizations, which offer advice on developmental issues such as education, governance, or economics. Sooner or later, such a society would adopt an organizational structure that resembles the structure of most states around the world.

Critically, we need to add, that it is equally conceivable that this society would now receive a special protected status—similar to a protected species—and only be accessible for scientific studies. After all, global models and ideas evolve and can change over time and it would depend on which cultural model becomes dominant—a development and progress-focused model or a model of independence, and non-intervention.

The Western Cultural Account

The backdrop for contemporary cultural models such as human rights and environmentalism is a specific 'Western cultural account' that is characterized by an emphasis on universalism, individualism, rationality, and so on (Meyer 2010; Schofer et al. 2012). Early roots of this account can be found in Christianity, but starting with the Enlightenment, humans have increasingly recognized that social life is not determined by a deity but can be influenced by humans. Consequently, ideas of progress and agency have developed. Initially, several models of modernity were available, but with the victory of the Allied Forces in World War II, cultural models have moved toward a 'more liberal, individualistic, and arguably "American" direction' (Schofer et al. 2012). The dominance of these values was further accelerated about 25 years ago by the breakdown of the communist system in Eastern Europe.

Of particular interest for the present study is the element of actorhood and agency that is prescribed in this cultural model. Even more, agency is not limited to personal matters. Rather, modern actors can claim agency and act on behalf of others and also on general ideas such as human rights, the environment, progress (Meyer 2010, 7). They act as disinterested 'Others' and in this role offer advice to other actors such as states, organizations, but also individuals. Consequently, an increasing emphasis on general principles and advising by impartial actors is observable, which has resulted in an increasing importance of science and rationalization principles. At the international level, these developments have resulted in the proliferation of organizations such as the United Nations, non-governmental organizations (NGOs), and professions that play an active role in the promotion of cultural models (see Boli and Thomas 1999).

This emphasis on disinterested others, however, does not mean that these actors are not involved in the diffusion of such cultural ideas. The modern social order is rather characterized by scriptwriting others, who prescribe agentic actorhood for actors, which can be seen in the enormous increase of educational efforts (Meyer 2010, 9). Education and curricula are increasingly influenced by global ideas and their contents have become more and more similar across countries. World Society scholars, thus, consider education an institution that increasingly links individuals and the global level.

An outcome of this actorhood focus is that individuals are perceived as being empowered, and also as being aware of their role as empowered agents and thus ready to act on these general principles. The dominance of this model of individual actors being the primordial actor (Meyer 2010, 8) has been accompanied by the demise of other models of modernity. Nationalistic doctrines of unified cultures that locate actorhood at the societal level have weakened, and ties such as family and community are now frequently depicted as based on individual choices. Consequently, human rights have been expanded globally more as human rights of individual persons (Soysal 1994) and less as rights of particular groups in a given society.

Organizational Sociology Concepts and Their Application in World Society

World Society is known for its emphasis on the global diffusion of cultural models. This research interest was initially sparked by the observation that schools in sub-Saharan Africa very much resemble schools in more affluent countries (Schofer et al. 2012). The conclusion was that such similarities cannot be explained easily by referring to functional needs and adoptions to local circumstances. Alternative explanations thus were needed.

Meyer and Rowan (1977) proposed that organizations and organizational structures reflect—and are aligned with—the cultural views of their environments and not merely shaped by technical requirements and processes. Meyer and Rowan used Berger and Luckmann's view (1966) of individuals acting and responding to a socially constructed world and applied it to the organizational realm. They further argue that an alignment with cultural views would affect not just a single organization, but would result in similarities across organizations.

This similarity in organizational forms and outcomes is termed 'isomorphism' and has become one of the initial foci of organizational sociology. At a global level, isomorphism can be observed as adoption of global ideas such as certain human rights or environmental standards in societies and organization around the world. Global institutions such as the United Nations, NGOs, and other active groups are perceived as playing an important role in the diffusion of such ideas or cultural models. They 'serve to institutionalize cultural models, effectively embodying and sustaining a global culture' (Schofer et al. 2012).

Isomorphism is only one side of the coin. Organizational studies are equally concerned with the spreading of ideas, the 'diffusion' of cultural models. Meyer (2010, 11) points out that realist models would refer back to individual interests when it comes to the proliferation of global ideas. The proliferation of certain austerity measures, for example, is explained by interventions of the International Monetary Fund and other groups. Such strong interests of dominant actors, however, are less able to explain the proliferation of human rights or environmentalism, as it is less obvious how and why dominant groups would win from the institutionalization of such principles. Here, cultural models and diffusion processes seem to be more appropriate.

World Society's emphasis on diffusion processes has led to the misperception that it predicts that everything will diffuse (Schofer et al. 2012). World Society, however, only asserts that ideas that emphasize collective goods over private interests, are aligned with global institutions, and are supported by international carriers will be likely to diffuse. Cultural models that do not have these three elements are less likely to diffuse. World Society related research was indeed able to identify such diffusion processes in areas such as the expansion of education, the rationalization of society, and the increasing importance of the individual and actorhood. Less consistent are the findings regarding human rights, where abuses are still abundant despite increasing human rights institutions (Schofer et al. 2012).

An important study in this regard is Boyle's (2002) work on the practice of female genital mutilation in Africa. Based on surveys and other sources, Boyle was able to show that the global dominant view of shunning this practice is reflected more in the attitudes of individuals, but less in their behaviors (Boyle et al. 2002). The connections and disconnections between the context and organizations as well as between policies and outcomes are discussed as coupling, loose coupling, decoupling, or re-coupling processes (see Bromley and Powell 2012). Applying this terminology to Boyle's study, coupling can be observed between global norms and attitudes, whereas loose coupling can be seen between global norms and behaviors, and also at the individual level between attitudes and behaviors.

Finally, World Society's emphasis on isomorphism has resulted in the misconception that it asserts that everything will become more similar. This critique notwithstanding, organizations are perceived as incoherent, lacking a common interest, drawn into different directions, and thus adopting contrarian views and cultural models, which in turn are frequently disconnected from the actual outcomes, particularly in resource-poor societies (Schofer et al. 2012). This view was already expressed in Meyer and Rowan's (1977) emphasis that organizations are influenced by contextual ideas, but are confronted with numerous, often contradictory, demands. Consequently, different parts of the same organization might work independently from each other and follow contrarian views, which can be kept up only by being disconnected, ceremonial management, and other symbolic work.

In sum, World Society applies concepts of organizational sociology such as isomorphism, diffusion, and various coupling and decoupling processes at a global level and studies how specific cultural models influence organizations and actors around the world. After summarizing these main concepts, the following section will now look into the typical empirical methods used by scholars working in this field.

Methods Used in World Society Related Research

World Society related assertions and questions have been studied in a variety of fields. This section thus focuses only on research that is relevant to the topics in this book, namely human rights and environmentalism. The following paragraphs thus discuss studies on the diffusion of human rights institutions (Koo and Ramirez 2009), decoupling between institutions and state behavior on human rights issues (Hafner-Burton and Tsutsui 2005), the proliferation of national parks and memberships (Frank et al. 2000a), and the origin and impacts of the global environmental regime (Hironaka 2014).

A central concern of World Society is diffusion processes, which call for a time-comparative approach. The most commonly used method is time-series analyses that capture the occurrence of different organizational outcomes in various countries around the world and link these changes to the influence of global cultural models. For example, Jeong-Woo Koo and Francisco Ramirez (2009) studied the adoption of two human rights institutions—the classic ombudsman and a human rights commission/ombudsman—between 1964 and 2004 in a total of 140 countries.

Koo and Ramirez recorded for each country-year observation whether such a human rights institution had been adopted or not, and showed that 104 out of 140 countries have adopted some form of these national human rights institutions. The occurrence of these adoptions is then linked to the influence of global ideas. Here, they considered both the global level, and ties between countries and the global level. The prevalence of and shifts in global ideas are grasped by the total number of international organizations as an indicator of the strength of the global civil society. In addition, they also developed a specific human rights indicator, considering the total number of international human rights instruments, world and regional adoption densities within a given year, and dummy variables for years where international conferences at the world level were held to grasp conference-related adoption behaviors. These variables are supplemented by indicators, which measure the links between nation-states and the global society. In this regard, Koo and

Ramirez consider the countries' memberships in INGOs and IGOs as well as the number of human rights treaties that have been ratified.

Koo and Ramirez focus on the diffusion process and attempt to capture both the influence of changing views within world society and the effects of the links between world society and organizations—in this case nation-states. Their results offer support for World Society, as all of the variables associated with this theory are significant in one or another model. The results, however, also show that neo-realist assumptions have an impact on the adoption of national human rights institutions and thus point to a parallel influence of alternative factors.

World Society also recognizes that not all ideas will diffuse and that not all adoptions will actually lead to changes in behavior. Decoupling processes between nation-state behavior and the adoption of human rights treaties were analyzed by Emilie Hafner-Burton and Kiyoteru Tsutsui (2005) in their analysis of ongoing state repression and human rights violations. Contrary to Koo and Ramirez (2009), Hafner-Burton and Tsutsui's study does not consider changes within the global world itself—only the link between countries and global ideas is measured by a country's membership in INGOs as well as by the ratification of various human rights treaties. Nevertheless, their results indicate that ratifications do actually have negative effects on repression and thus make matters worse, whereas ties via the global civil society (INGOs) can have positive effects.

Hafner-Burton and Tsutsui conclude that some states may sign treaties as mere 'window dressing' and do not change their behaviors. This would be a typical case of radical decoupling. Later studies, however, also point out those negative effects could be caused by better recording practices after the ratification of treaties as well as that these initial negative effects can be offset over time, as NGOs and activities can use these treaties as leverage to make governments move in a certain direction (see Cole 2012b).

Similar models and approaches have also been applied in environmental studies. A frequently cited study is Frank, Hironaka and Evans' (2000a) work on the influence of World Society on the nation-state and the natural environment over the twentieth century. Using an event analysis, their study looks into the causes of the proliferation of national parks, chapters of environmental associations, memberships in intergovernmental organizations, and environmental impact assessment laws and environmental ministries globally between 1900 and 1988.

Their dependent variables capture the occurrence of these different environmental institutions at the national level. A first set of independent variables considers the global level of environmentalism by capturing the staff size of the United Nations Environment Programme, a dummy variable for the years with UN environment conferences, and the total number of international treaties on the environment. A second set of variables considers the ties between the nation-state and World Society and includes the national chapters of international non-governmental associations and national memberships in international governmental organizations—excluding environmental chapters and organizations. Finally, Frank et al. also include national receptor sites—the number of national ecology associations and the number of natural science organizations, respectively.

Frank et al.'s (2000a) analyses provide support for World Society, given that variables associated with this theory were more influential than variables associated with alternative explanations such as the domestic level of environmental degradation or the national economic affluence. However, Buttel (2000) criticized this analysis for not considering hard environmental outcomes—the difference between ceremony and substance discussion. Frank et al. (2000b) acknowledged in their reply that the environmental regimes and treaties are far from perfect. Yet, they also see some instances of improvement such as the water quality of many rivers in more developed countries, which led them to the conclusion that 'even a pockmarked system is better than no system at all' (Frank et al. 2000b, 123).

These three studies represent the typical approach of World Society related research, using time-series and event analyses. Considering their independent variables, the frequent use of the variables 'number of International Governmental Organizations (IGO)' and 'number of International Non-Governmental Organizations (INGO)' is striking and has led to the misperception within the scientific community of equating World Society with an INGO effect (Schofer et al. 2012). The underlying understanding, however, rather seems to be that ties via INGOs serve also as a proxy for the links between nation-states and global ideas, whereas the actual flow of ideas can happen in various ways. Hironaka (2014) thus proposed a 'Bee Swarm' model of social change, in which numerous forces such as international treaties, IGOs, INGOs, laws and policies, and others factors work simultaneously, which makes it difficult to track down the influence of a single factor. Furthermore, capturing the effects of global ideas becomes even more difficult once global models have become institutionalized and thus are embedded in various entities such as organizations, laws, curricula.

Regardless of these conceptualization issues, empirical studies still need to capture the tangible effects of global ideas, and various INGO, IGO, and international treaty measurements seem to be the indicators of choice. Even Ann Hironaka (2014), who proposes the Bee Swarm model in her book on Greening the Globe, has added an appendix with various regression models on pollutants such as CO_2 and SO_2; protected land areas, which use environmental treaties; and the known INGO variable as independent variables.

In sum, these different studies demonstrate that World Society related research considers three main levels. Firstly, the global level and dominant cultural models; secondly, the link between nation-states and the global level via INGOs, IGOs, and various treaties; and thirdly, national receptor sites such as local environmental chapters. The next section will show that similar concepts are also applied to the effects of global ideas on individuals.

Furthermore, albeit not discussed in detail in this chapter, all approaches also consider alternative explanations and theories, such as realist approaches by Koo and Ramirez (2009) and the domestic level of environmental degradation or the national economic affluence by Frank et al. (2000a). The empirical chapters of this book thus will also briefly discuss alternative views that are important when it comes to environmental behaviors and attitudes toward human rights.

Notes

1. Parts of this chapter were published in Hadler (2015).

References and Further Readings

Berger, P.L., and T. Luckmann. 1966. *The Social Construction of Reality: A Treatise in the Sociology of Knowledge*. Garden City, NY: First Anchor.

Boli, J., and G.M. Thomas., eds. 1999. *Constructing World Culture: International Nongovernmental Organizations Since 1875*. Stanford, CA: Stanford University Press.

Boyle, E.H. 2002. *Female Genital Cutting: Cultural Conflict in the Global Community*. New York: The Johns Hopkins University Press.

Bromley, P., and W. Powell. 2012. From Smoke and Mirrors to Walking the Talk: Decoupling in the Contemporary World. *The Academy of Management Annals* 6: 483–530.

Buttel, F.H. 2000. World Society, the Nation-State, and Environmental Protection. *American Sociological Review* 65 (1): 117–121.

Coase, R. 1937. The Nature of the Firm. *Economica* 4: 386–405.

DiMaggio, P.J., and W.W. Powell. 1991. Introduction. In *The New Institutionalism in Organizational Analysis*, ed. W.W. Powell and P.J. DiMaggio, 1–38. Chicago, IL: University of Chicago Press.

Frank, D.J., A. Hironaka, and E. Schofer. 2000a. The Nation-State and the Natural Environment over the Twentieth Century. *American Sociological Review* 65 (1): 96–116.

———. 2000b. Environmentalism as a Global Institution: Reply to Buttel. *American Sociological Review* 65 (1): 122–127.

Greenwood, R., C. Oliver, R. Suddaby, and K. Sahlin-Andersson, eds. 2008. *The SAGE Handbook of Organizational Institutionalism*. London: Sage.

Hadler, M. 2015. Institutionalism and Neo-Institutionalism: History of the Concepts. In *International Encyclopedia of the Social and Behavioral Sciences*, ed. J.D. Wright, vol. 12, 2nd ed., 186–189. Encyclopaedia Entry.

Hafner-Burton, E., and K. Tsutsui. 2005. Human Rights Practices in a Globalizing World: The Paradox of Empty Promises. *American Journal of Sociology* 110: 1373–1411.

Hall, P.A., and R.C. Taylor. 1996. Political Science and the Three Institutionalisms. *Political Studies* 44: 936–957.

Hironaka, A. 2014. *Greening the Globe. World Society and Environmental Change*. Cambridge: Cambridge University Press.

Immergut, E.M. 2011. Institution/Institutionalism. In *International Encyclopedia of Political Science*, ed. B. Badie, D. Berg-Schlosser, and L. Morlino. London: Sage.

Jepperson, R.L. 1991. Institutions, Institutional Effects, and Institutionalism. In *The New Institutionalism in Organizational Analysis*, ed. W.W. Powell and P.J. DiMaggio, 143–163. Chicago, IL: University of Chicago Press.

Koo, J.-W., and F. Ramirez. 2009. National Incorporation of Global Human Rights: Worldwide Expansion of National Human Rights Institutions 1966–2004. *Social Forces* 87 (3): 1321–1353.

Menard, C., and M.M. Shirley, eds. 2005. *Handbook of New Institutional Economics*. Dordrecht: Springer.

Meyer, J.W. 2010. World Society, Institutional Theories, and the Actor. *Annual Review of Sociology* 36: 1–20.

Meyer, J.W., J. Boli, G.M. Thomas, and F.O. Ramirez. 1997. World Society and the Nation-State. *American Journal of Sociology* 103: 144–181.

Meyer, J.W., and B. Rowan. 1977. Institutionalized Organizations: Formal Structure as Myth and Ceremony. *American Journal of Sociology* 83 (2): 340–363.

Nelson, R., and S.G. Winter. 1982. *An Evolutionary Theory of Economic Change*. Cambridge: Harvard University Press.

North, D.C. 1991. Institutions. *The Journal of Economic Perspectives* 5: 97–112.

Powell, W.W., and P.J. DiMaggio, eds. 1991. *The New Institutionalism in Organizational Analysis*. Chicago, IL: University of Chicago Press.

Schmidt, V.A. 2008. Discursive Institutionalism: The Explanatory Power of Ideas and Discourse. *Annual Review of Political Science* 11: 303–326.

Schofer, E., A. Hironaka, D. Frank, and W. Longhofer. 2012. Sociological Institutionalism and World Society. In *The New Blackwell Companion to Political Sociology*, ed. E. Amenta, K. Nash, and A. Scott, 57–68. Oxford: Wiley-Blackwell.

Scott, W.R. 2014. *Institutions and Organizations: Ideas. Interests, and Identities*. 4th ed. Thousand Oaks, CA: Sage.

Selznick, P. 1949. *TVA and the Grass Roots: A Study in the Sociology of Formal Organization*. Berkeley: University of California Press.

Senge, K. 2007. Was ist neu am Neo-Institutionalismus. *Österreichische Zeitschrift für Soziologie* 32: 42–65.

Skocpol, T. 1979. *States and Social Revolutions: A Comparative Analysis of France, Russia and China*. Cambridge: Cambridge University Press.

Soysal, Y.N. 1994. *Limits of Citizenship: Migrants and Postnational Membership in Europe*. Chicago, IL: The University of Chicago Press.

Veblen, T. 1899. *The Theory of the Leisure Class: An Economic Study of Institutions*. New York: Macmillan.

Williamson, O.E. 1975. *Markets and Hierarchies, Analysis and Antitrust Implications: A Study in the Economics of Internal Organization*. New York: Free Press.

3

World Society and the Individual

As pointed out in Chap. 2, actorhood is one of the dominant themes of World Society's Western cultural account. This can be seen in narratives regarding the individual impact on social change and the emphasis on individual choice in many contemporary writings. Yet, while this picture of an agentic actor is central to modern polities—especially in the liberal polities, such as in the USA—the different neo-institutionalisms and related theories differ in the way an actor is described.

A very general line can be drawn between theories that follow realist premises and those that follow phenomenological ideas (Meyer 2010; Schofer et al. 2012). Realist approaches consider individual actors, their interests, and so on, as naturally given and outside the institutional framework, whereas phenomenological approaches perceive them as derivatives of the institutional structure. 'In phenomenological models, actors are constructed entities, playing parts as in the theater' (Meyer 2010, 4). Furthermore, realist approaches consider the institutional environment as quite narrow, with a rule-like character, and often limited to a single institution such as property rights, whereas phenomenological approaches consider the institutional environment as complex cultural meanings (Meyer 2010).

M. Hadler, *The Influence of Global Ideas on Environmentalism and Human Rights*,
DOI 10.1057/978-1-137-57440-4_3

This first divide between realist and phenomenological approaches can be extended. Jepperson (1991) offers a more detailed classification based on two distinct dimensions—(a) the degree to which actors are socially constructed and (b) the featured level of analysis. The first category distinguishes phenomenological approaches with their social constructivist view from realist approaches. The second category distinguishes methodologically individualistic approaches from structuralist views. The cross-tabulation of these two dimensions results in four distinctive groups that represent different theories and approaches.

The first group includes sociological institutionalisms such as World Society that combine phenomenological thoughts with a structuralist focus. The second one contains approaches such as neo-institutional economics, which are characterized by individualistic and realist views, and thus represent the opposite to World Society in both dimensions. The third group refers to approaches such as organizational culture or Weick's phenomenologically inspired social psychology. These approaches share the social-constructivist view with World Society, but focus on the construction of individuals and are thus also of interest for the present study. Finally, the fourth group includes approaches such as Social Ecology, which follow a bottom-up paradigm and focus on structural aspects, based on individualistic methodological considerations. This group thus is less relevant for studying influences on individuals.

Jepperson (1991) did not mention the research on institutional entrepreneurs. This research focuses on individuals who invent new institutions or combine existing institutions in novel ways. They need to assume that an actor's interests are partially external to the institutional structure and are thus close to realist approaches, which does not bode well for the phenomenological core assumption, which considers the actor's interests and identities as elements of the institutional framework (Hallett 2010, 67). In terms of Jepperson's overview, these entrepreneurial approaches are thus located between the economic neo-institutionalism and approaches such as organizational culture.

Before discussing these approaches in more detail, it is also worth noting that these theories also differ in their views on the direction of the influences (Powell and Colyvas 2008). One group emphasizes how individuals develop, maintain, and through their actions change institutions—a bottom-up approach—whereas another group focuses on how individuals are constricted by certain institutions. World Society is rather in line with the second view—given its focus on the diffusion of global ideas and how local actors change accordingly. However, as Powell and Colyvas (2008) point out, this process can be also recursive. While this recursive relationship is recognized in the literature, empirical studies—as will be discussed later in this chapter—focus most often only on a single direction.

3.1 The Individual in World Society

World Society related research has a strong focus on the influence of global ideas on organizations and organizational outcomes, which has led to the critique that the impact of institutions on individual actors has not been acknowledged (see Hallett 2010). At the center of this critique lies the observation that sociological neo-institutionalism has started out by making strong references to micro-sociological concepts and by applying micro-sociological insights to organizations and organizational behavior. These micro-sociological accounts, however, have lost influence over time—a development called 'macro-evolutionary drift' by Hallett (2010).

Further, the argument about the influence of institutions on individuals, according to Hallett, reverses the initial concepts. 'While institutions were originally viewed as human social constructions, organizations and people are increasingly treated as institutional constructions (Meyer and Jepperson 2000)' (Hallett 2010, 56). Given this development, it is helpful to briefly revisit the micro-sociological roots of World Society before discussing the contemporary attempts of reconnecting individual-level aspects to World Society. Meyer and Rowan's (1977) seminal piece on formal structures as myth and ceremony makes direct references to P. Berger and T. Luckmann's *The Social Construction of Reality* (1966) and Goffman's *Interaction Rituals* (1967). Further, as Powell and Colyvas (2008) point out, Garfinkel's *Studies in Ethnomethodology* (1967) was important in the foundational writings of sociological neo-institutionalism.

When explaining the core idea of isomorphism—that organizational structures reflect contextual ideas, named 'myths,' and that increasing alignments with these myths will ultimately result in an increasing similarity across organizations—Meyer and Rowan (1977) use Berger and Luckmann's (1966) views and state that organizations and their structures reflect socially constructed reality (and are not only caused and shaped by technical interdependencies and requirements.)

In addition, Berger and Luckmann's (1966) views on the institutionalization process is also of particular relevance. They perceive it as a three-step process of externalization, objectification, and internalization. Ideas or new behaviors, initially, need to be expressed and made external. Later, they become shared knowledge and thus objective facts, which in turn are internalized by individuals. Powell and Colyvas (2008, 280) summarize: 'As particular schemas become routinized through repeated application ... they develop a habitual, taken-for-granted character.' In this process, 'local' elements become 'natural' and are recognized as guideposts (ibid., 292).

Meyer and Rowan (1977) admit that adherence to different myths can be problematic for organizations, as they may impede efficiency, cause contradictory demands, and lead to inconsistent structures. In resolving such inconsistencies, organizational structures can decouple structures, but still maintain an image of a well-run enterprise. Here, Meyer and Rowan use Goffman's (1967) idea of maintaining face and the three associated practices of avoidance, discretion, and overlooking. Powell and Colyvas (2008, 279) summarize this as organizations using a public and a backstage face in order to cope with different demands from the organizational environment. They also add that Goffman's views on interaction rituals link the individual interaction rituals to the larger social order, which is also a core assumption in sociological institutionalism.

While not directly quoted in their seminal piece, Powell and Colyvas (2008) also emphasize that Garfinkel's (1967) studies on ethnomethodology informed the developing sociological institutionalism. Ethnomethodology had a strong focus on organizations and the workplace. It contends that categories and classifications become interpretative schemas, which in turn become routinized if used frequently, and taken-for-granted accounts. Organizations, in this view, cannot be separated from the individuals, as will be discussed below when describing Lynne Zucker's (1977) work on the maintenance of institutions.

Conceptualization of the Micro Level

Research on the influence of institutions on actors refers commonly to Ronald Jepperson (1991), who describes actors as being constituted by institutions, with institutions defining which actions are perceived, and considered thinkable at all. The interests of actors and their choices are part of the institutional environment and actors are likened to 'stage actors' who follow certain scripts (Meyer 2009). The term 'actor' is used rather loosely and can refer to organizations and individuals. Meyer's (2010) contribution on world society and the actor, for example, refers to nation-states, schools, organizations, and individuals equally—similar logics presumably apply to all of them.

The core assumption that actors—individuals and organizations—are influenced by institutions goes so far that those institutions are seen as defining the actors' goals and identities (DiMaggio and Powell 1983; Meyer and Rowan 1977). Given the various uses of the term 'institution' as pointed out in the previous chapter, Jepperson (1991, 149) defines institutions 'as socially constructed, routine-reproduced (ceteris paribus), program or rule systems. They operate as relative fixtures of constraining environments and are accompanied by taken-for-granted accounts.'

The taken-for-granted aspect of institutions and also of their content refers to the fact that they do not need be reinforced by actions and constant interventions (Jepperson 1991). It, however, must not be likened with unawareness. Actors, as Jepperson emphasizes, can follow rules regardless if they are aware of it or not. Some actors might even intensively study and scrutinize certain rules and yet follow them. Further, institutions have two sides—they are simultaneously enabling and constraining actions. Compulsory voting, for example, constrains behavior—as it forces individuals to cast their vote—but at the same time enables citizens to take part in the political process of a country.

As for the substantive content of institutions, several terms are commonly used. Meyer and Rowan (1977) used the term 'myth' and defined it as 'rationalized and impersonal prescriptions that identify various social purposes as technical ones and specify in a rule-like way the appropriate means to pursue these technical purposes rationally.' They consider myths as 'highly institutionalized and thus is in some measure beyond the discretion of any individual participant or organization. They must, therefore, be taken for granted as legitimate …' (Meyer and Rowan 1991, 44).

Equally, terms such as 'typifications,' 'schema,' 'routines,' 'scripts' are used to refer to similar aspects. Scott (2014, 96) provides an overview of these terms and assigns them to different carriers and pillars of institutions. As for the cultural cognitive pillar, which is emphasized by World Society, the terms 'categories,' 'typifications,' 'schemas,' and 'frames' refer to *symbolic carriers* such as language. The terms 'predispositions' and 'scripts' refer to *activities* that carry and reflect institutions. The terms 'isomorphism' and 'identities' are used with regard to *relational systems* and the term 'objects' (that possess symbolic value) is used with regard to *artifacts*.

Other institutional pillars that are less emphasized in World Society are the regulative and the normative ones (Scott 2014). The former includes carriers such as laws, government and power systems, monitoring and sanctioning activities, as well as objects that fulfill mandated specifications. The latter includes carriers such as values and standards, regimes and authority systems, routines and habits, as well as artifacts that meet certain standards.

Independent from Scott's classification, Jepperson (1991) perceives three primary carriers—formal organizations, regimes, and culture. Formal organizations refer to specific entities such as work organizations. Regimes relate to a central authority and codified rules, which are sanctioned. Culture, finally, refers to customary and rather conventional institutions, which are not associated with any formal organizations or central authority. In addition, individuals themselves are also possible carriers of global ideas (Schofer et al. 2012)—an aspect that becomes more important given the increasing status of actorhood in global models (Frank and Meyer 2002; Suárez 2008).

Regardless of the differences in these classifications and emphases, research following the mainstream World Society approach of using macro-level time-series analyses frequently includes international organizations as a main indicator for links to global models. This has led to the misconception of equating World Societal effects with the International Non-Governmental Organizations (INGO) effect (Schofer et al. 2012). With a view on the different carriers, these INGOs should be rather considered a proxy that measures the links between a society and global ideas, the distribution of cultural cognitive ideas, and thus the influence of various carriers.

When Do Institutions Matter?

The detailed classifications of institutional carriers and pillars of institutions seem to allow for close examination of various aspects of institutional influences. However, World Society and its phenomenological approach faces some obstacles as pointed out by John Meyer (2010) using the example of the myth of actorhood.

The myth of actorhood is a core element of the Western Cultural account (see Chap. 2) and has increased tremendously over the last centuries (Meyer 2010; Schofer et al. 2012). As for an individual, these historical developments are reflected in the shifts in the societal views of what a peasant, a citizen, and an actor are supposed to do. Peasants are supposed to be treated humanely, but are not expected to make any demands. Citizens of modernity are supposed to have interests, may seek to influence authorities, and are allowed to make demands, but do not necessarily act on behalf of common interests. The contemporary actor, however, '… presumably has authorized agency not only to represent self-interest, but also to choose interests and even actively to manage the rules of the social environment' (Meyer 2010, 3).

Yet, in a phenomenological approach that considers actors as following scripts, empirical analyses face the problem that the modern dominant cultural model of an agentic individual has led to individuals that display (and recount) purposive behaviors (Meyer 2010, 5). Meyer points in this regard to the example that many college students have a ready narrative of why, how, and when they decided to attend college, whereas this decision was actually never a choice, but rather the scripted path to be taken. Actions and 'decisions,' however, are often described post hoc as rational choice, which causes the problem for the phenomenological institutionalism that many scripted actions appear as if they were rational choices. In contrast to realist models, in which the conformity results from rational reasoning and goal pursuing behaviors (e.g. education for higher income), World Society theory sees it rather through unconscious processes, given that actors are constituted by institutions.

Furthermore, as Hironaka (2014) emphasizes, it is difficult to pinpoint an actual causal mechanism and to underscore one specific factor, when it comes to the influence of global ideas. Research that attempts to do this, in her view, follows a 'smoking gun' model, whereas she is proposing a 'Bee Swarm' model. 'In a Bee Swarm model of social change, a dense network of causal factors shapes outcomes, even if most individual causal factors prove weak or inconsequential' (Hironaka 2014, 7). Despite these difficulties, World Society's emphasis on cultural models has allowed various insights by studying the underlying cultural models that constitute actors and drive change. Ignoring interests and power—Schofer et al. (2012)—has opened the possibility for studying the influences of global ideas on processes such as the decolonization process in Africa, the proliferation of human rights, and environmentalism—all processes that cannot be easily be explained by references to the interests of specific powerful groups.

The underlying consensus regarding the substantial impact of cultural models on the identities of actors—modern persons, groups, and societies—is that the impact of these external scripts is the strongest 'among those actors best linked to the wider environment, through worlds of associational life (e.g. non-governmental memberships) and cultural life (e.g., through extended schooling)' (Meyer 2010, 13). This also applies to organizational outcomes, when considering Schofer and Hironaka's (2005) description of how and when nation-states are more likely to adopt world societal ideas: The more international organizations are present, the longer they have been present, and the more contact points between the international and national community exists.

Processes of decoupling, however, are quite likely, given the numerous external principles and the sheer impossibility for an actor to follow and comply with all expectations (Meyer 2010). A diffuse character of the modern actor is thus the norm, as seen in Boyle et al.'s (2002) study on female genital mutilation in Africa. Individual attitudes were aligned with the international conventions implemented in these countries, whereas behaviors did not follow suit. We thus can see a (known) gap between attitudes and behaviors, yet in terms of institutional language decoupling processes would be present between global ideas and behaviors. With an increasing number of demands, the opportunities for decoupling will increase. Meyer (2010) reverses the direction of this thought and concludes in his overview that the many inconsistencies between the numerous expectations and the limited ability of actors to follow all expectations will allow for plenty of growth for institutions that promote these models, given that any state, organization, individual, and the like, will be non-compliant in some regards (Meyer 2010, 14).

World Society scholars, in addition, also recognize the influence of national institutions and policies on how the myth of individual actorhood is implemented. In this regard, Jepperson and Meyer (1991) have developed a categorization of modern polities and distinguish first between polities that consider individual actors as carriers of public and private tasks and polities that clearly distinguish between social (private) and public activities. Secondly, they consider the ways political systems deal with different tasks and distinguish between centrally organized polities with a tight control of sociopolitical processes and polities that consider sociopolitical processes as a natural outgrowth of a society.

These two dimensions lead to four different categories of how individuals are supposed to act as public agents: (1) liberal/individualist polities such as the USA with a strong emphasis on individual action and a less centralized state; (2) statist societies such as France with a centrally organized state and an emphasis on an empowered individual; (3) corporatist polities such as Austria with a centrally organized state and a clear distinction between public and social activities; and (4) segmental societies such as most of the Latin American countries with the state being outside the society.

Relevance for This Book

Applying these views to the topics of the present book, individuals can be considered as constituted by cultural models that contain views on human rights and environmentalism. Examples of such cultural models are the rise of the ecosystem model, which connects individuals and the environment (Schofer et al. 2012). Global ideas of environmental issues thus might inform and shape everyday actions of individuals, in whose lives labels such as 'organic' or 'BPA-free' now have a specific significance

(Hironaka 2014, 16). The same principle can be applied to human rights, where labels such as Fair Trade also convey certain meanings.

These cultural models will be diffused in various ways, with World Society emphasizing the cultural cognitive pillar and—in its empirical studies—the role of international organizations. There are, however, also other pillars of institutions, namely the normative and regulative ones (Scott 2014). These pillars are reflected in certain standards such ISO 14000 on environmental management procedures or laws about recycling programs and fines for illegal dumping of waste. Empirical studies, ideally, should consider various possible carriers.

In addition, cross-national differences can also occur due to different polity demands on individual participation, in line with the classification of Jepperson and Meyer (1991). Its usefulness becomes visible when considering patterns of environmental protest. In a European context, radical actions are more frequent in France—a society characterized by a polity that encourages civil action—than in other European countries (Kriesi et al. 1995). The development of resistance to genetically modified crops in France and in corporatist Austria illustrates this difference (Seifert 2008). In Austria, a consensus between the government and NGOs was achieved and implemented through laws. In France, activists used means such as the burning of crop fields and subsequent court trials to influence public opinion and political leaders.

These examples also underscore the phenomenological assumption that an individual can be, but does not need to be aware of institutional demands, but is constituted by them regardless. A French farmer who is engaged in a dismantling of a McDonald's restaurant—such as Bovi in 2001—most likely will not explain such actions in terms of living in a polity that encourages civil actions, and, yet, be more active than comparable social groups in other countries. Similarly, an individual that considers humans and the environment as interwoven will most likely not point to the changes in science in the twentieth century, which has led to the rise of this view.

Overall, in line with the assumptions that global ideas more likely to influence individuals and organizations the better they are linked to global cultural models (Meyer 2010; Schofer and Hironaka 2005), exposure to carriers such as educational systems or residency in areas with more international organizations seem to be crucial factors that need to be captured empirically. Yet, other studies and theories point to a more active role of the individual that goes beyond the mere exposure idea. Some of these approaches will be discussed in the next section.

3.2 The Individual in Other Approaches

This section summarizes approaches that explicitly consider individual-level processes and actors. In terms of Jepperson's (1991) typology, the focus is on approaches that share the phenomenological view with World Society, such as Lynne Zucker's early work and Weick's sensemaking approach, and also on concepts such as institutional entrepreneurs, which are more in line with economic neo-institutionalism. Furthermore, the more recent development of 'inhabited institutions' (Hallett 2010; Hallett and Ventresca 2006) as well as the framing approach to social movements (Benford and Snow 2000) will be addressed.

In the same year as Meyer and Rowan's piece on myth and ceremony, Lynne Zucker—a PhD student of John Meyer, published the article 'The role of institutionalization in cultural persistence' in the *American Sociological Review* (Zucker 1977). Zucker combines the phenomenological view with an ethnomethodological approach and by investigating three different aspects of institutionalization—transmission, maintenance, and resistance to change. Her analysis focused on acts that are highly institutionalized and thus both objective and exterior to an individual. Objective, in this view, refers to being potentially repeatable by others and exterior to an existing intersubjective understanding. Zucker tested the assumption that social knowledge, once institutionalized, is able to influence behaviors without being internalized, rewarded, or otherwise intervened. For this purpose, an experiment was set up, in which participants needed to assess the movements of a stable light source in a dark room.

Institutionalization was varied by calling the confederate—who announced an initial assessment of the movements—'another person,' 'a co-worker,' and 'the light operator.' These three types are supposed to reflect increasing levels of institutionalization. 'As predicted, it was found that the greater the degree of institutionalization, the greater the generational uniformity of cultural understandings, the greater the maintenance without direct social control, and the greater the resistance to change through personal influence' (Zucker 1977, 742). The learned assessment (the judgment of the confederate) was more likely to be upheld in more institutionalized situations, i.e. when co-workers or the light operator made the call, compared to setups in which 'another person' made the call.

Zucker, in this early work, answered the question of how and why institutions are maintained (Scott 2014, 151). Later, she turned toward the question of institutional entrepreneurs in biotech and other industries. Together with Pamela Tolbert (Tolbert and Zucker 1999), she offered a view on the actual institutionalization processes in organizations pointing out how single actors or individuals react to changes in the environment and come up with new ideas, of which a few might turn out to be useful and are taken up by other actors. Once others adopt a specific act, a habitualization has occurred. Further, the benefits of the action might get theorized and thus enter general knowledge—the objectification of the action. Sedimentation is the final step, when the new action spreads throughout an entire population and becomes a new standard (Tolbert and Zucker 1999, in Scott 2014, 148f).

The focus on entrepreneurs can be related to DiMaggio's (1988) ideas on interest and agency in institutional theory. Institutional entrepreneurs were also considered by Neil Fligstein (2001) who added that individuals have social skills and can use them to alter institutional content, create new meanings, and deal with diverse interests and various individuals. However, as Hallett (2010, 67) points out, this picture of an active individual that follows a personal interest in developing new meanings is quite inconsistent with the basic sociological neo-institutionalism view that interests are considered part of the institutional environment.

Other related approaches are sensemaking and status expectation research (Powell and Colyvas 2008). Karl Weick's (1995) research on sensemaking focused on the question of how individuals enact social order, by considering their identity and their social context. '... individuals act based on who they are, not on what choices they have, and this feature is constituted out of the process of interaction' (Powell and Colyvas 2008,

284). Sensemaking is primarily a retrospective process and occurs particularly when individuals confront unexpected situations. In these cases, 'individuals reach into their repertoire of experience to make a situation fit the immediate circumstances and allow them to resume their work' (Powell and Colyvas 2008, 293). Insofar, there is a strong overlap with the example John Meyer (2010) discussed when students recount the reasons why they 'decided' to go to college.

Powell and Colyvas (2008) also mention research on status expectation. This line of research points out that the interpersonal status within groups is often assigned based on general stereotypes and thus connects wider societal beliefs and views with individual actions and everyday activities. This research resonates with Zucker's (1977) experiments, which also showed that the status of the confederate—another person, a co-worker, or the light operator—influences the likelihood of not changing one's initial view on the movement of the light source.

Whereas the sensemaking and status expectation research exist independently from sociological neo-institutionalism, the inhabited institutions approach (Hallett and Ventresca 2006; Hallett 2010) has developed in direct reaction to this sociological approach. Revisiting and re-interpreting Gouldner's classic study *Patterns of Industrial Bureaucracy*, Hallett and Ventresca's propose an inhabited institutions approach that considers the embeddedness of institutions and interactions, and differences in their meanings, which can range from the very local to the broad and public level, and, methodologically, follows a skeptical and inquiring attitude. Their main emphasis is that institutions are not inert, but enacted by individuals in interactions, which in turn can also change institutions.

Their proposed open inquiring research attitude is not limited to a specific research methodology. Hallett and Ventresca (2006), at the same time, suggest ethnography as the most suitable approach, as it is easier for an ethnographer to consider the local context, the actual interactions, and the embedding in broader cultural meanings and constraints. This approach was later empirically applied in Hallett's (2010) study on the re-coupling of a local school to the broader macro-cultural myth 'accountability.' Hallett shows how a newly appointed principal attempts to implement accountability. His study highlights local meaning processes, differences between the staff and the new principal, and the conflict-laden implementation process—and thus show that the enactment of broader institutions is not conflict-free, but rather a battleground.

Hallett (2010) tries to overcome the strong empirical focus on macro-level phenomena and the poor conceptualization of individual-level processes in sociological neo-institutionalisms such as World Society. The inhabited institutions approach is considered complementary to macro-level approaches, which have the tendency to depict the individual as enacting myths in a rote way. It also presents an alternative to institutional entrepreneur approaches, which depict actors as independent from the institutions, a view that does not bode well with the phenomenological view that actors and interests are constituted by the institutional framework (Hallett 2010, 67).

Hallett and Ventresca (2006) as well as Scott (2014) point to recent development in social movement literature and the emphasis of some scholars such as Benford and Snow (2000) on framing processes. This strand of social movement literature shares many similarities with neo-institutional research, but replaces the term 'wider institutional logics' with the term 'field frames.' Erving Goffman developed the concept of framing, almost a decade later than his interaction rituals, which had inspired Meyer and Rowan (1977). Benford and Snow (2000, 6) apply the concept of frames to social movements and point out that it is the shared understandings which define whom to blame for a particular issue and influence others to participate in or influence change. They define these particular tasks as 'diagnostic framing,' 'prognostic framing,' and 'motivational framing.'

Benford and Snow (2000) define diagnostic framing as the act in which a problem is identified and possible causes are blamed. Direct action, in their view, is very much dependent on the identification of possible causes. The second framing task is prognostic framing. This is where a social movement expresses a resolution for a specific issue. This involves the formulation of possible approaches for putting their plan into action. However, it is this step that separates social movements from one another. Similar movements can have the same diagnostic framing, but their approaches can differ greatly. Thirdly, motivational framing, the last task in framing, is what offers a reason for taking part in collective action. This includes the construction of movement-specific language and terms. It is with a new vocabulary that adherents are provided with motivation to participate in collective action and a reason to sustain it.

This brief overview of other approaches that also are concerned with the influence of cultural models on individual actors shows that these views add an additional emphasis on the active role of the individual not only in the process of institutionalization, but also in changing and altering existing institutions and cultural models. They thus go beyond the focus on exposure, which was dominant in the previous section. The final section of this chapter now turns to empirical studies and discusses how these different views and approaches have been used in international comparative research.

3.3 The Individual in Empirical Studies

This final section focuses on empirical research that attempts to connect individual attitudes and behaviors to global cultural models by explicitly referring to World Society theory. This overview, however, is limited to studies that use individual-level outcomes based on survey research—the approach that is used in this book as well. In addition, it emphasizes the two main areas of this book: environmentalism and human rights. All studies are briefly introduced, and this chapter concludes with a brief summary of their merits and shortcomings as well as with an outlook on the empirical approach of the present study.

Studies on Environmental Behavior and Attitudes

In a previous paper with Max Haller, I have studied (Hadler and Haller 2011) the influence of world societal ties on the individual private and public environmental behavior using the 2000 ISSP data on environmental attitudes. A total of 23 countries were included and ties to World Society were measured by the number of governmental and non-governmental organizations present in a given society. The polity contexts as a proxy of different demands on individual participation—as discussed in this chapter—were included as well. Using a multi-level approach, we were able to show various influences of IGOs and INGOs. INGOs, however, had a stronger influence on private behavior than on public behavior. The analyses did not consider any interactions between ties to World Society and other variables and thus follows a simple exposure approach—along the lines that the more organizations are present in a society, the stronger is their impact.

When the International Social Survey Programme published the 2010 environment data, the cross-sectional study (Hadler and Haller 2011) was extended to a longitudinal analysis for the public behavior dimension by considering all three waves of ISSP data (Hadler 2016). This extended study considered 16 countries. Ties to world society were captured by the number of INGOs, IGOs, and environmental INGOs (EINGOs) present in a given country. The findings suggest a positive influence of the overall number of NGOs, but negative effects of their growth in a given county. In addition, this study also goes beyond mere exposure models, by considering an interaction between individual education and INGOs, which shows that international ties and education influence each other.

Sandra Marquart-Pyatt (2012) studied the influence of various contextual influences on three environmental concerns using 27 countries from the 2000 ISSP survey. Dependent variables are (a) environmental threat awareness that consists of questions asking about how dangerous various pollutants are for the environment, (b) environmental efficacy measured by the subjective views of respondents on how important is it to act, and (c) the respondents' willingness to pay for environmental protection. Ties to world society are measured by the countries' memberships in environmental INGOs. Other macro variables include social affluence, industrialization, democracy, and pollution indicators. Her analyses show that INGOs are important in most dimensions. Their effect, however, changes when the individual-level effect of education is allowed to vary. Thus, whereas Marquart-Pyatt follows a simple exposure model, the education dependency of the INGO, however, points to a mediating role of education.

Kyle Knight and Benjamin Messer (2012) use four waves of World Value Survey (WVS) data to study the association between individual environmental concern, societal affluence, environmental degradation, and ties to world society in a total of 85 countries. The dependent variables are aggregated individual-level responses to five dimensions of environmental concerns: willingness to pay, local seriousness, global seriousness, environmental protections, and confidence in environmental movements. Knight and Messer thus consider more dimensions, countries, and time points than other studies. Controlling for GDP and pollution, their analyses do not indicate any significant influences of the world societal ties—which are measured as the density and the number of INGOs in a given country. They, however, limit their study to aggregate values, which does not allow the study of any micro-level or cross-level effects.

Finally, several pieces on the influence of world societal ties on environmental outcomes were published by scholars working with Andrew Jorgenson. Individual concerns are studied in Givens and Jorgenson (2013) and Jorgenson and Givens (2014). Givens and Jorgenson's studies include data from 37 countries from the fifth wave of WVS data. Concern is measured by an item that asks respondents to compare themselves to descriptions of hypothetical persons. Givens and Jorgenson used the item to describe a person, for whom it is important to care for nature. Ties to world society are measured by a per capita measure—the number of EINGOs in a given country, divided by the population size. These ties, as well as the presence of an environmental ministry show positive effects on environmental concern, also when controlled for other macro-level factors. Givens and Jorgenson (2013) see the individual as constituted by the world polity and thus are also part of the exposure models. Their twin paper Jorgenson and Givens (2014) includes a few additional variables, but shows basically the same results.

Studies on Matters of Human Rights

The field of human rights covers a wide range of topics. The studies summarized here thus cover specific topics such as tolerance toward certain groups, general views toward the locus of human rights power, as well as specific practices such as genital mutilation. All studies, however, share a focus on individual responses, the use of survey data, and a special concern with the influence of world societal ties.

Elizabeth Boyle, Barbara McMorris, and Mayra Gomez (2002) studied the local conformity to international laws when scrutinizing individual attitudes and behaviors with regard to female genital mutilation. Using survey data from five African countries, they find that institutions such as education, mass media, and female employment—all carriers of modern scripts—have a significant lowering impact on women's approval of such practices. Modernization-related factors, such as the availability of electricity, had positive effects on the attitudes, but no effects on behavior. They suggest that attitudes might change before behavior (2002, 22). Links to global ideas, via mass media, education, and employment, had positive effects on both attitudes and behaviors. At the same time, membership in social groups that might oppose global ideas showed that Islamic and women of traditional African belief were more likely to circumcise their daughters and also more in favor of continuing this practice. Boyle et al. (2002) include several interactions effects between carriers of international scripts and individual characteristics. Interestingly, compared to other studies, exposure to international norms was measured using items such as individual education, but also having a radio or working outside home. This study is thus a more direct approach than the simple national exposure models that include only the number of ties to World Society at the country level. On the other hand, however, the actual links between a society and global ideas are not considered.

In an article on homophobia and xenophobia (Hadler 2012), I have used WVS data to study the effects of world societal ties on individual tolerance toward different social groups. Respondents were asked if they mind living next to foreigners, individuals with a different religious background, homosexuals, and the like. These items were, then, categorized in various underlying dimensions and the determinants of homophobia and xenophobia studied in more detail. A multi-level regression including three waves of WVS survey data from 32 different countries indicated positive effects of international ties—the number of INGOs and IGOs in a given country. These variables, however, were included only at level two—they thus measure the effects of increasing numbers of organizations, but do not fully capture variation across countries. This is reflected only in varying intercepts. The analyses did not consider any interactions between INGOs and individual-level variables and thus can be described as an 'exposure' approach.

Min Zhou (2013) uses WVS data from the 5th wave, covering 49 countries, when analyzing the responses to the question whether 'human rights policies should be decided by the national governments, by regional institutions, or by the United Nations.' Ties to world society were included as INGO and IGO, as well as number of international treaties that have been signed by a given country. In addition, he recognizes education as a main carrier of global ideas and thus includes an interaction with world societal ties, which turned out to be significant. His conclusions are that international ties matter, but not the signed treaties. The acceptance of an international approach to human rights is thus higher in countries with more ties to world society and accelerated by education, even when controlled for other macro-level factors such as democratization and the level of development. His analysis thus considers different carriers of global ideas—INGOs and treaties—and also goes beyond the simple expose models by considering interaction effects between ties to world society and education.

Rachael Pierotti (2013) studies the changes in attitudes toward intimate partner violence through the lenses of World Society and demography. Using two waves of the Demographic and Health Survey from 26 low- to middle-income countries, Pierotti tested if changes in attitudes follow the global ideational change toward a rejection of partner violence and if these changes remain significant once controlled for demographic shifts such as increasing urbanization, educational access, and media access as well as within and between cohort exchanges.

Pierotti did not use direct measures of world societal links at the macro level as she analyzed each country separately. Her findings are nonetheless in line with World Society assumptions, as a sudden and significant change toward the rejection of partner violence occurred within the cohorts and thus cannot be explained by cohort replacement. Her findings also support the idea that education, urban residency, as well as media access have positive effects and thus might be possible carriers of global cultural models. Interestingly, younger cohorts were less likely to reject partner violence, which is not in line with the idea of a stronger influence of global ideas on younger respondents.

Ann Velitchkova (2015) analyzed Afrobarometer data from 15 sub-Saharan countries and studied the influence of the respondents' micro-level institutional embedding in family, ethnic group, and religious background on the use of political violence. She locates her research as an institutional micro-foundation approach and argues that embedding in micro-level institutions that oppose global world societal views will increase the likelihood of violent political behavior. She considers two types of institutional carriers—practices and self-identification. These carriers are captured by survey questions such as active religious practices and the personal importance of religion, respectively. Macro-institutional differences such as the presence of INGOs are discussed and considered important in the selection of the countries. The analysis itself, however, includes only dummy variables for each country.

Velitchkova's work thus can be compared to the approach of Boyle et al. (2002) when studying female genital mutilation. In contrast to Boyle et al., Velitchkova makes references to the inhabited institutions approach of Hallett and Ventresca (2006) and frames her analyses in clear micro-level terms. Novel in her approach is to include practice measures, which reflect behaviors that are opposed to global norms such as experiencing violence in one one's family or expecting favoritism toward one's family or group. Such micro-level logics contradict the global norms propelled by international institutions and convey a less homogeneous depiction of the social world than a diffusion model.

Common Themes and Differences

The studies described in this section can be characterized along a few distinctive dimensions—the inclusion of nation-global links via multi-level approaches versus separate studies of single countries or sets of countries, an exposure approach versus a more inhabited institutions approach, time-comparative versus cross-sectional studies, and the consideration of changes at the global level versus the consideration purely of links between nations and the global level.

The difference between multi-level models and studies that consider countries separately is decisive when considering the links between countries and global ideas. Multi-level models as used by Marquart-Pyatt (2012), Zhou (2013), Jorgenson and Givens (2014) that combine national public opinion data with country-level characteristics such as the national affluence and links to world society are able to empirically test the effects of these ties on the opinions in different countries. Studies such as Boyle et al. (2002), Pierotti (2013), and Velitchkova (2015) that analyze survey data from different countries in separate models reveal more details about intra-country differences and effects, but cannot grasp the actual effects of the ties to world society.

All of these studies, however, include to varying degrees an exposure approach, as they assume that the context of a respondent, such as the country of residence and its links to world society, or residency in urban or rural areas, has an influence on their attitudes and behaviors, without actually being able to qualitatively describe the nature of this context or how ideas from this context diffuse to an individual. In a similar vein, demographic variables such as age and education are only proxies, with the underlying idea that global ideas are more prevalent in recent times and thus have a stronger effect on younger respondents and are more prevalent in higher levels of education.

These exposure ideas have been expanded in two ways. Firstly, multi-level studies such as Zhou (2013) and Hadler (2016) consider interactions between education and the presence of international organizations in different countries, along the line that international ideas are more successfully included in curricula the more ties to the global world exist. Secondly, Velitchkova (2015) goes beyond an exposure approach by including behavioral and membership aspects that contradict global models and thus measure congruency and discrepancies between individual behaviors, attitudes, and social embedding and dominant global models.

Another important aspect is if these models use cross-sectional or longitudinal data. World Society theory tries to capture diffusion processes of global ideas over time. Most studies on organizational outcomes thus follow a time-comparative approach (see Chap. 2). By contrast, the survey studies discussed in this section use predominantly cross-sectional data—also due to the limited availability of longitudinal material. Cross-sectional studies such as Hadler and Haller (2011) or Zhou (2013) can only present a snap shot in time and cannot grasp diffusion processes, whereas longitudinal cross-sectional studies can differentiate between a cross-cultural and growth effect as will be pointed in more detail in the empirical chapters of this book.

Alongside this longitudinal versus cross-sectional question, the studies mentioned in this section mostly ignore the changes at the world societal level itself. Global ideas change over time and can differ between regions. Measuring the links to world society in terms of numbers of organizations captures only the links, but not the changes at the global level. Capturing the changes at the global level was done in Koo and Ramirez's study on human rights institutions (see Chap. 2) by considering the global changes via cumulative numbers of organizations, densities of organizations, and other factors.

In sum, these studies using survey data represent an intermediate level between the ethnography approach of Hallett (2010) and the dominant macro-level diffusion models of World Society discussed in Chap. 2. The consideration of different countries allows capturing and testing the effects of different contexts, which are missed in a single ethnography, whereas the inclusion of individual characteristics allows for capturing embedding aspects that are missed in macro-level approaches. On the other hand, such survey-data-based studies are less detailed than ethnographies and usually include fewer countries than the organizational outcome studies due to limited data availability.

The empirical chapters in this book attempt to address several of these shortcomings by using time-comparative and cross-national survey data and including a variety of contextual and individual-level variables. Furthermore, by contrasting two topics—human rights and environmental behaviors—this book will also address another shortcoming of many of the aforementioned studies, a focus on a single topic. This seems to be particularly relevant as the overview of studies on the environment, for example, has shown that world societal links and cultural ideas can have different effects on similar topics. The examination of different topics attempted in this study thus will allow for deeper insights.

References and Further Readings

Benford, R.D., and D.A. Snow. 2000. Framing Processes and Social Movements: An Overview and Assessment. *Annual Review of Sociology* 26 (1): 611–639.

Berger, P.L., and T. Luckmann. 1966. *The Social Construction of Reality: A Treatise in the Sociology of Knowledge*. Garden City, NY: First Anchor.

Boyle, E.H., B.J. McMorris, and M. Gomez. 2002. Local Conformity to International Norms: The Case of Female Genital Cutting. *International Sociology* 17 (1): 5–33.

DiMaggio, P.J. 1988. Interest and Agency in Institutional Theory. *Institutional Patterns and Organizations: Culture and Environment* 1: 3–22.

DiMaggio, P.J., and W.W. Powell. 1983. The Iron Cage Revisited: Collective Rationality and Institutional Isomorphism in Organizational Fields. *American Sociological Review* 48 (2): 147–160.

Fligstein, N. 2001. Social Skill and the Theory of Fields. *Sociological Theory* 19 (2): 105–125.

Frank, D.J., and J.W. Meyer. 2002. The Profusion of Individual Roles and Identities in the Post-War Period. *Sociological Theory* 20: 86–105.

Garfinkel, H. 1967. *Studies in Ethnomethodology*. Englewood Cliffs, NJ: Prentice Hall.

Givens, J., and A.K. Jorgenson. 2013. Individual Environmental Concern in the World Polity: A Multilevel Analysis. *Social Science Research* 42: 418–431.

Goffman, E. 1967. *Interaction Ritual: Essays in Face to Face Behavior*. Chicago, IL: Aldine.

Hadler, M. 2012. The Influence of World Societal Forces on Social Tolerance. A Time Comparative Study of Individual Prejudices in 32 Countries. *The Sociological Quarterly* 53: 211–237.

———. 2016. Individual Action, World Society, and Environmental Change: 1993–2010. *European Journal of Political and Cultural Sociology* 3 (2–3): 341–374.

Hadler, M., and M. Haller. 2011. Global Activism and Nationally Driven Recycling: The Influence of World Society and National Contexts on Public and Private Environmental Behavior. *International Sociology* 26: 315–345.

Hallett, T. 2010. The Myth Incarnate. Recoupling Processes, Turmoil, and Inhabited Institutions in an Urban Elementary School. *American Sociological Review* 75: 52–74.

Hallett, T., and M.J. Ventresca. 2006. Inhabited Institutions: Social Interactions and Organizational Forms in Gouldner's Patterns of Industrial Bureaucracy. *Theory and Society* 35 (2): 213–236.

Hironaka, A. 2014. *Greening the Globe. World Society and Environmental Change*. Cambridge: Cambridge University Press.

Jepperson, R.L. 1991. Institutions, Institutional Effects, and Institutionalism. In *The New Institutionalism in Organizational Analysis*, ed. W.W. Powell and P.J. DiMaggio, 143–163. Chicago, IL: University of Chicago Press.

Jepperson, R., and J. Meyer. 1991. The Public Order and the Construction of Formal Organizations. In *The New Institutionalism in Organizational Analysis*, ed. W.W. Powell and P.J. DiMaggio, 204–231. Chicago, IL: University of Chicago Press.

Jorgenson, A.K., and J.E. Givens. 2014. Economic Globalization and Environmental Concern: A Multilevel Analysis of Individuals within 37 Nations. *Environment and Behavior* 46 (7): 848–871.

Knight, K., and B. Messer. 2012. Environmental Concern in Cross-National Perspective: The Effects of Affluence, Environmental Degradation, and World Society. *Social Science Quarterly* 93: 521–537.

Kriesi, H., R. Koopmans, J.W. Duyvendak, and M. Giugni. 1995. *New Social Movements in Western Europe: A Comparative Analysis*. London: UCL Press.

Marquart-Pyatt, S.T. 2012. Contextual Influence on Environmental Concern Cross-Nationally: A Multilevel Investigation. *Social Science Research* 41: 1085–1099.

Meyer, J.W. 2009. Reflections: Institutional Theory and World Society. In *World Society: The Writings of John W. Meyer*, ed. G. Krucken and G.S. Drori, 36–63. Oxford: Oxford University Press.

———. 2010. World Society, Institutional Theories, and the Actor. *Annual Review of Sociology* 36: 1–20.

Meyer, J.W., and R.L. Jepperson. 2000. The 'Actors' of Modern Society: The Cultural Construction of Social Agency. *Sociological Theory* 18: 100–120.

Meyer, J.W., and B. Rowan. 1977. Institutionalized Organizations: Formal Structure as Myth and Ceremony. *American Journal of Sociology* 83 (2): 340–363.

———. 1991. Institutionalized Organizations. Formal Structure as Myth and Ceremony. In *The New Institutionalism in Organizational Analysis*, ed. W.W. Powell and P.J. DiMaggio, 41–62. Chicago, IL: University of Chicago Press.

Pierotti, R. 2013. Increasing Rejection of Intimate Partner Violence. Evidence of Global Cultural Diffusion. *American Sociological Review* 78 (2): 240–265.

Powell, W.W., and J.A. Colyvas. 2008. Microfoundations of Institutional Theory. In *The SAGE Handbook of Organizational Institutionalism*, ed. R. Greenwood, C. Oliver, R. Suddaby, and K. Sahlin-Andersson, 276–298. London: Sage.

Schofer, E., and A. Hironaka. 2005. The Effects of World Society on Environmental Protection Outcomes. *Social Forces* 84: 25–47.

Schofer, E., A. Hironaka, D. Frank, and W. Longhofer. 2012. Sociological Institutionalism and World Society. In *The New Blackwell Companion to Political Sociology*, ed. E. Amenta, K. Nash, and A. Scott, 57–68. Oxford: Wiley-Blackwell.

Scott, W.R. 2014. *Institutions and Organizations: Ideas. Interests, and Identities*. 4th ed. Thousand Oaks, CA: Sage.

Seifert, F. 2008. Einhellige Abschottung und globalisierungskritische Bauernrevolte. Widerstand gegen die grüne Gentechnik in Österreich und Frankreich. *SWS-Rundschau* 48 (4): 485–504.

Suárez, D.F. 2008. Rewriting Citizenship? Civic Education in Costa Rica and Argentina. *Comparative Education Review* 44: 485–503.

Tolbert, P.S., and L.G. Zucker. 1999. The Institutionalization of Institutional Theory. In *Studying Organization. Theory and Method*, ed. S.R. Clegg and C. Hardy, 169–184. London, Thousand Oaks, CA, and New Delhi: Sage.

Velitchkova, A. 2015. World Culture, Uncoupling, Institutional Logics, and Recoupling: Practices and Self-Identification as Institutional Microfoundations of Political Violence. *Sociological Forum* 30 (3): 698–720.

Weick, K.E. 1995. *Sensemaking in Organizations*. Vol. 3. Thousand Oaks, CA: Sage.

Zhou, M. 2013. Public Support for International Human Rights Institutions: A Cross-National and Multilevel Analysis. *Sociological Forum* 28 (3): 525–548.

Zucker, L.G. 1977. The Role of Institutionalization in Cultural Persistence. *American Sociological Review* 42 (5): 726–743.

4

The International Level: IGOs and INGOs

The previous chapters pointed out that World Society scholars emphasize the role of the international community and frequently consider international governmental organizations (IGOs) and international non-governmental organizations (INGOs) in their empirical research. These indicators, however, are just proxies for the influence of global cultural models and cannot reflect the many different organizations and complexity of this international level.

This chapter now introduces a few main IGOs and INGOs with regard to human rights and environmentalism. In technical terms, the Union of International Associations (2017) defines IGOs as 'an organization composed primarily of sovereign states, or of other intergovernmental organizations. IGOs are established by treaty or other agreement that acts as a charter creating the group. Examples include the United Nations, the World Bank, or the European Union.' INGOs, on the other hand, are defined as follows. 'The United Nations Economic and Social Council (ECOSOC) defines an INGO as "any organization which is not established by inter-governmental agreement (Resolution 288 (X) 27 February 1950), including organizations which accept members designated by

M. Hadler, *The Influence of Global Ideas on Environmentalism and Human Rights*,
DOI 10.1057/978-1-137-57440-4_4

government authorities, provided that such membership does not interfere with the free expression of views of the organizations" (Resolution 1296 (XLV) of 25 June 1968)' (ibid.).

IGOs and INGOs thus cover a wide range of different organizations. Ann Hironaka (2014) likens the complexity and the effects of the international level to a bee swarm. The overall impact of the swarm can be seen and tested—using the proxies of IGO and INGO—but the specific effects of the many single actors and organizations might be difficult to grasp. Yet, contrary to this view, there are international players such as the United Nations (UN), which have devised instruments with tangible effects—such as treaties on emission reductions and covenants on human rights. The first part of this chapter thus introduces the UN, emphasizing its human rights and environment foci.

The non-governmental sector is even more diverse than the governmental sector and the NGOs discussed in the second part of this chapter are no more than examples that highlight certain trends. Sierra Club, Friends of the Earth, and Greenpeace are introduced to show different organizational structures and approaches to action in environmentalism. Amnesty International (AI), Human Rights Watch (HRW), and Oxfam are used to do the same for human rights.

This chapter focuses only on one aspect of international developments. Other forces (see Harper 2012) such as international corporations and the associated economic globalization as well as international trade organizations (Stiglitz 2002), historical cleavages in the world-system (Wallerstein 2004), and cultural aspects such as global flows of people, ideas, and media (Appadurai 1996) that may also influence attitudes and behaviors are not discussed here. The empirical chapters of this book (Chaps. 6 and 7), however, consider such alternative approaches and include associated indicators in the analyses.

4.1 The United Nations

The UN is the prime example of an IGO. The UN evolved after World War II as the successor to the League of Nations, which failed to prevent this war (see Hanhimäki 2015; Kennedy 2006; Meisler 1995). The

League of Nations was riddled with problems due to its emphasis on unanimous decisions, collective security, and limited geographic scope. The UN was thus sketched out during World War II and became effective in October 1945, when the five permanent members—the USA, France, Great Britain, China, and Russia—and several of the regular members signed the UN Charter. The main goal of the UN is maintaining peace and international security, but also promoting human rights, social and economic development, humanitarian aid, and protecting the environment. The present summary focuses only on the aspects of human rights and environmental protection.

The UN and Human Rights

The issue of human rights has been addressed throughout modern history, but as Kennedy (2006, 178f) points out, the UN approach was more encompassing than all previous efforts, which were often limited to treatment during wars, such as the Hague Conventions of 1899 and 1907. However, while the initial charter emphasized the importance of human rights, it also states explicitly that none of its content authorizes the UN to intervene in matters that are within domestic restrictions. In terms of organization, the United Nations Commission on Human Rights (UNCHR) was established in 1946 by the United Nations Economic and Social Council (ECOSOC) to deal with human rights. As of 2006, the UNCHR was replaced by the United Nations Human Rights Council (HRC), which works closely with the Office of the High Commissioner for Human Rights (OHCHR)—an office that was established in the wake of the Human Rights conference in 1993.

The core document of the UN on human rights is the Universal Declaration of Human Rights, adopted by the UN on 10 December 1948. It goes back to a drafting committee, which was established after World War II by the UNCHR. The declaration covers the right to life, liberty, and security, rights in civil society, political rights, and social security and well-being. It passed the general assembly in 1948, despite quarrels in the background due to different positions and concerns of various countries, such as colonial powers, racial discriminations, and socialist regimes. Of course, many states assumed that it would not be implemented as it was put forward by the General Assembly and thus had no binding power.

Yet this declaration started a process of implementation, monitoring, and measuring (Kennedy 2006, 180f.). Some states, such as the Scandinavian countries, took the declaration at face value and implemented instruments that helped to achieve the goals of this declaration. Regardless, the content of this declaration offered a publicly available measure against which the different actions and programs of governments could be measured. The declaration thus gave rise to a civil society, which repeatedly emphasized the goals of this declaration. It thus developed power and instigated pressure to meet human rights goals despite the lack of any implicit sanctions within this declaration.

The pressure from civil society is particularly important, as actions by the UN on behalf of human rights are hampered by the structure of the UN. The UN Commission on Human Rights reports to the ECOSOC and thus to the General Assembly. Given that the General Assembly cannot enforce action, the central roles of the UNCHR are advising and reporting. Only the Security Council can sanction interventions, by statute. Yet the council is often in a stalemate due to the different political positions of the USA, Russia, and China.

Besides its reporting task, the UN Human Rights Commission was also asked to create a body of international human rights laws and conventions based on the declaration, and to devise mechanisms needed to enforce its implementation (see www.ohchr.org). Two main outcomes of this task are the International Covenant on Civil and Political Rights and the International Covenant on Economic, Social and Cultural Rights, which were adopted in 1966 and came into force in 1976. In parallel, other treaties were developed[1] such as the Convention on the Prevention and Punishment of the Crime of Genocide (adopted in 1948, came into force in 1951) and the Convention Relating to the Status of Refugees and Protocol Relating to the Status of Refugees (adopted in 1951, came into force in 1954), and the Convention on the Elimination of All Forms of Racial Discrimination.

Contrary to the UN human rights declaration, these covenants become binding for countries that ratify them and individuals whose human rights are violated could appeal in court. Many countries thus do not accept the whole agreement and have exemptions. For example, the USA is not willing to restrict any domestic rights to freedom of speech obligated through the Convention on the Elimination of All Forms of Racial Discrimination: The USA stated upon signature: 'The Constitution of the United States contains provisions for the protection of individual rights, such as the right of free speech, and nothing in the Convention shall be deemed to require or to authorize legislation or other action by the United States of America incompatible with the provisions of the Constitution of the United States of America' (UN Treaty collection 2016).

International covenants and human rights issues are debated at specific conferences. World Society theorists point to the accelerating power of international conferences, which result in increased activities before conferences, when states are trying to catch up with various conventions and after the conferences when states are ratifying certain documents (Hironaka 2014; Koo and Ramirez 2009). As for human rights, the UN has organized two large conferences with the term 'human rights' in their name—the 1968 Tehran conference at the 20th anniversary of the human rights declaration and the 1993 World Conference on Human Rights in Vienna.

The first human rights conference was hosted in Tehran, Iran—in regime of the Shah, which was not democratic and not particularly pro human rights (Burke 2008). Furthermore, this conference took place during the Cold War period, where the UN had more of an observing role due to the mutual veto of the super powers in the Security Council. UN interventions thus were an exception and investigation, data collection, and so on, were more important. As an outcome, this conference only urged countries to enforce human rights principles and was considered a failure, according to Burke.

The political climate started to change in the 1970s and 1980s, with rapid democratization developments in Africa and Latin America, the collapse of the socialist system in Eastern Europe and the USSR, and the end of the apartheid regime in South Africa. The changes increased the relevance of human rights topics, and another global conference was organized in 1993 in Vienna (OHCHR 2017). The conference brought together governmental and non-governmental organizations, including grassroot movements, and attracted around 7000 participants. The outcome of this conference was the Vienna Declaration and Program of Action, which led to the formation of the High Commissioner on Human Rights and a call to strengthening human rights around the globe.

Yet, as Kennedy (2006) points out, civil society associations were disappointed with the developments at the Viennese conference, because they were treated as bystanders without full inclusion. Furthermore, some of the most atrocious human rights violations such as war crimes in the former Yugoslavia, in Ruanda, and so on, coincided with this conference and offered a rather bleak view on the state of the human rights.

Other important conferences are the World Summit 2005 in New York City and the series of World Conferences against Racism. The World Summit led to the foundation of the HRC, given that the predecessor organization, the UNCHR, was criticized heavily for including countries with poor human rights records. HRC member countries are now elected by the general assembly and the general assembly may 'by a two-thirds majority of the members present and voting, … suspend the rights of membership in the Council of a member of the Council that commits gross and systematic violations of human rights' (UN Durban Review 2017).

Among the series of World Conferences against Racism, the 2009 conference in Geneva and 2011 follow-up meeting in New York are of special interest (UN Durban Review 2017). The Geneva meeting was summoned to review the implementation of the Durban Declaration and Programme of Action from the 2001 conference on the same topic. The 2009 meeting, however, was overshadowed by the participation of the Iranian President Mahmoud Ahmadinejad and his speech on Israel and the Holocaust, which resulted in the boycott of this conference, and also of its follow-up conference in 2011 by several countries, including the USA, Canada, Germany, Austria, and the Czech Republic. In substance, the conference's outcome was a condemnation of people due to their ethnic, religious, or other minority statuses.

The UN and the Environment

Environmental problems and environmental pollution, in contrast to domestic human rights violation, can have a direct impact on other countries as seen in the issues of nuclear fallout, ozone depletion, and climate change. On the other hand, as described in the tragedy of the commons (Hardin 1968), states and actors are less likely to reduce their environmental pollution when they fear that other countries will not reduce their pollution output. The solution of such problems often requires good international coordination, as can be seen in the more frequent international conferences on this topic.

Yet, environmental topics have been taken up by the UN much later than Human Rights. The first major conference on environmental issues took place in 1972 in Stockholm (Hironaka 2014). Before that, environmental topics were addressed loosely by different institutions, such as air noise by an aviation agency, forestation by forestry departments, and so on. The occurrence of the Stockholm conference, however, was initially uncertain, as environmental protection was considered a topic of the affluent societies, whereas developing countries face 'real' problems such as economic progress, life quality, and so on.

The conference became a success, only after environmental issues were combined with development issues (Hironaka 2014). Nowadays, the Stockholm conference is considered the beginning of an international environmental policy. An organizational outcome of the Stockholm conference was the founding of the United Nations Environment Programme (UNEP). The UNEP, with its headquarters in Nairobi, Kenya, deals with a wide range of environmental issues, including assisting developing countries in sustainable development and establishing of environment-friendly policies. The UNEP continues the successful legacy of the Stockholm conference, by continuing to merge the topics of environmental problems and development issues (Hironaka 2014).

Alongside the UNEP, the UN has several agencies and organizations that deal with environment-related issues. Among the oldest international organizations is the World Meteorological Organization (WMO), having its roots in the International Meteorological Association, which was established in 1873. When the issue of climate change became more salient, it led to the founding of the Intergovernmental Panel on Climate Change (IPCC) in 1988. Its tasks are the review and compilation of scientific research results on climate change. Interestingly, the US Reagan administrators were some of the main proponents of this IGO, as they favored an organization that combines scientific information and governmental overview (Weart 2011).

Environmental Treaties

Considering two main global environmental problems of the recent past, the depletion of the ozone layer and climate change, what can be said about attention to these problems at the international level? To answer this question, the Montreal Protocol and the treaties that have been agreed upon within the United Nations Framework Convention on Climate Change (UNFCCC) are discussed in the following paragraphs.

The depletion of the ozone layer became a salient issue in the 1980s, when evidence was found that the ozone layer on top of Antarctica was depleted. This offered strong support for earlier laboratory research that suggested the possibility of such effects (see Harper 2012, 64ff). After scientific research had identified a few chemicals as culprits, concerted international action was agreed on. First, 'The Vienna Convention for the Protection of the Ozone Layer' provided a framework in 1985 and specific targets were agreed upon in the Montreal Protocol.

The Montreal Protocol was signed in 1987 and became effective in 1989 (Gareau 2013; Grundmann 2001). A total of 197 parties ratified this protocol, with both the European Union and each member country of the European Union being a member (UNEP). It thus has a much wider agreement than the human rights treaties such as the International Covenant on Civil and Political Rights, which has been ratified only by 169 countries (http://indicators.ohchr.org/). The environmental impact of the subsequent actions and phasing out of CFCs can be seen in the fact that the ozone layer on top of the Antarctic is slowly recovering. Furthermore, the Montreal Protocol is unique, due to its geographical scope, its effectiveness, and evidence-based policy (Canan et al. 2015).

Yet, as Harper (2012, 67) points out, 'the Montreal protocol is a risky predictor for how quickly and effectively other international negotiations may turn out,' given that only a few gases had to be banned and only a few companies were concerned, whereas other problems may require much wider action. This assessment is true when considering the most pressing issue currently—climate change, or more precisely anthropogenic-induced climate change.

Climate change has been observed for a while, given that changing weather patterns and temperatures were already noticed more than 150 years ago. Yet only in 1979, did the WMO organize a meeting on climate in Geneva (Harper 2012, 68). A formal UN institutionalization occurred within the UNFCCC (UNFCCC 2017), which was agreed upon during the 1992 Rio de Janeiro conference—the Earth Summit. Currently, the UNFCCC has 197 parties and thus a near-universal membership.

The UNFCCC offers only a general framework for treaties and negotiations, but does not call for certain emission reductions, legal requirements, or enforcement mechanisms. This is possibly one of the main reasons for the near-universal membership, which also includes the USA and Canada—two countries that did not participate (USA) or withdrew (Canada) from the associated Kyoto treaty.

Members of this treaty have annual meetings—the Conferences of Parties (COP)—and have devised two major international agreements within the UNFCCC—the Kyoto Protocol and the Paris Agreement. The former asked for firm commitments of its members and included enforcement mechanisms, whereas the latter follows a bottom-up approach and builds upon voluntarily actions of its member states and relies on a naming and shaming approach. Considering the World Society emphasis on cultural institutions and less so on enforced principles, the Paris Agreement is more in line with this theory.

The Kyoto Protocol (Kyoto Protocol 1998) was agreed upon with the goal to reduce the emission of greenhouse gases (GHG) to 'a level that would prevent dangerous anthropogenic interference with the climate system' (Art. 2). It was adopted in Kyoto, Japan, in 1997 and became effective in 2005, when Russia ratified it and pushed the treaty over the '55 countries—55% of the Greenhouse Gas emissions' enforcement threshold. The protocol has two commitment periods, with the first one from 2008 to 2012. At the COP meeting in Doha, a second period from 2012 onwards was negotiated. This second period, however, is not yet active, as too few countries have ratified.

The initial problem with the Kyoto Protocol was that it did not include all major emitters of GHG. Developing countries were not obligated to reduce their emissions and developed countries such as the USA (and later Canada) refused to ratify the protocol (or left the agreement later on as in the Canadian case). Despite these limitations, a number of countries signed and ratified the treaty and committed themselves to reducing their GHG output.

Several countries were able to reduce their output. This success, however, was partially also due to the breakdown of the socialist economies in Eastern Europe and Russia, which led to huge decreases in emissions in these countries. Given that Kyoto also allowed for emission trading systems, these countries were in the comfortable situation of first, having reduced emissions due to their collapsed economies, second, the room to re-expanding their economies without too much worry about environmental impacts, and third to trade the remaining certificates.

The Kyoto Protocol enforced the commitments in the following way. Countries that made pledges to reduce their GHG emissions had to reduce them to the level of 1990 (and in some cases to an earlier base-year). If these countries failed to achieve their commitment, it would have to catch up in the 2nd period. Yet, it would have to save an additional 30% and would not be allowed to use trading schemes anymore. Several countries had to buy certificates to meet their initial pledges in order to avoid penalty payments. Canada, facing substantial fines, on the other hand, decided to opt out of the Kyoto Protocol in 2011.

The second period—after 2012—was addressed at the 2012 COP meeting in Doha. Given the many problems with the first round, it is not surprising that the second round was less of a success. The outcome was an agreement to extend Kyoto to 2020 and to work on a successor program, with 2015 as a target date. As of now, only 37 countries have binding Kyoto II targets, and the successor program—the Paris Agreement—is based on voluntary commitments.

In 2015, at the COP meeting in Paris, the Paris Agreement was achieved (Paris Agreement 2015), which is supposed to deal with climate change from 2020 onwards. The main goals of the Paris Agreement are to keep the temperature increase well below 2C, to foster a nation's ability to adapt to climate change, and to enhance sustainable development (Paris Agreement, Art. 5). Yet, contrary to the Kyoto Protocol, the Paris

Agreement does not impose any internationally determined emission reduction targets or sanctions for states.

The Paris Agreement became effective in 2016. It was ratified in 144 countries, and currently has 197 parties. It is thus more accepted than the Kyoto Protocol. One reason, as mentioned above, is that it relies on voluntary actions of its member countries. Each country needs to determine a national level of contribution to GHG reduction, which will be monitored, but does not lead to sanctions if it is not met. This voluntary system allowed also the USA to take part, as the Byrd-Hagel Resolution did not apply. Due to the lack of enforcement, the USA considers it rather an executive agreement and not an international treaty, therefore the US Senate did not have to be consulted (Taraska 2015).

Every second year, the countries need to report their progress, which will be reviewed and assessed by experts and peers. Contrary to Kyoto, countries are not grouped into different annexes with and without commitments; rather the level of ability is considered in reporting requirements and contributions efforts. Similar to Kyoto, countries can work together and exchange contributions. Every five years, the overall effort of all countries and the progress toward the overall goal are assessed. This overall assessment, based on initial contributions, will take place for the first time in 2018. The agreement anticipates a stepwise process, with the national contributions becoming progressively greater.

4.2 Regional Forces and Counter Trends

The description of the UN institutions and major INGOs emphasized the pro-environment and a pro-human rights views and institutions. Yet, there are also counter forces, as can be seen in the frequent stalemates in the UN Security Council between the USA, Russia, and China when it comes to interventions, or the reluctance of several countries to sign and ratify international agreements such as the Kyoto Protocol. This final section thus highlights a few international main cleavages that can be seen with regard to human rights and the environment.

Regional Forces and Counter Trends in Human Rights

The Universal Declaration of Human Rights claims universal applicability, yet various countries contested it after it was drafted and signed it only because they considered it ineffective (Kennedy 2006). This perception came from its location within the General Assembly and not within the Security Council, which can call for interventions.

Over time, voices became even more diverse, as an increasing number of former colonies became independent and raised their own concerns.

The increasing power of these former colonies was seen in the adoption of the Convention on the Elimination of All Forms of Racial Discrimination and particularly at the first international conference on human rights in 1968, in Tehran, where it became obvious that the universal scope of human rights is clearly in doubt (Burke 2008). For the new independent countries—often with autocratic systems, such as in Africa, but also in the Middle East, Latin America, and Asia—economic development was more important than individual human rights. For development, a 'concentrated state power was lauded as the path to freedom' (Burke 2008, 296).

The Tehran conference thus marked a shift from individual rights to national development. The Universal Declaration of Human Rights was considered a Western project, reflecting ideas of the past that do not fit the current state of the world. As Burke (2008) points out, even the two superpowers, the USA and the USSR, came closer in their positions at this conference, as they were equally concerned about the power of the new nations.

More recently, the Organization of the Islamic Conference adopted the Cairo Declaration on Human Rights in Islam (1990), which aims to combine religious principles and human rights. Examples in this regard are the declaration stating that 'all men are equal in terms of basic human dignity and basic obligations and responsibilities, without any discrimination on the grounds of race, colour, language, sex, religious belief, political affiliation, social status or other considerations' (Art. 1a, Cairo Declaration).

Yet, it also gives more weight to men than women, as the husband is considered responsible for the support of the family (Art 6b), and gives precedent to the Shari'ah in articles such as 'There shall be no crime or punishment except as provided for in the Shari'ah' (Art 19d) or that 'Everyone shall have the right to express his opinion freely in such manner as would not be contrary to the principles of the Shari'ah' (Art 22a). These differences have also affected the positions of member countries on various issues. As Chap. 7 will show, the topic of tolerance toward homosexuality is particularly affected and the UN HRC meeting has seen boycotts and walkouts in regard to this issue.

These examples highlight some counter forces. Yet, the opposite is also observable, where the UN declaration has evoked regional institutions. In Europe, for example, the European Convention on Human Rights was adopted in 1953 by the Council of Europe, which is composed of 47 European community member states. This body was formed to strengthen human rights and promote democracy and the rule of law. Individuals whose human rights were infringed upon can appeal to the European Court of Human Rights in Strasbourg, but must have attempted all domestic means.

Regional Forces in Environmentalism

The discussion of the nature of human rights around the Tehran conference pointed to concerns about economic development in less affluent countries. The same cleavage also applies to environmentalism. During this period, environmentalism was considered a luxury, which only rich societies can afford, and as Hironaka (2014) points out, only the inclusion of the development issue made it possible to organize the first major conference on environmental topics in Stockholm in 1972.

The relationship between developed and less developed countries is also of importance with regard to environmental impact and discussed within the literature on economic exchange (Jorgenson et al. 2011; Shandra et al. 2009). This research points out that developed countries are able to offshore and outsource their environmental problems, related to basic industry such as production and mining. A monetarization of this exchange has occurred in the various trading schemes, which allows countries to sell their pollution shares—which turned out to be one of the reasons why Russia signed the Kyoto Protocol in 2004.[2]

Drawing a line between developed and less developed countries does not grasp the full picture. Considering the Kyoto Protocol, lines can be drawn between developed countries as well. Kyoto was not ratified by the USA as the Senate passed the Byrd-Hagel Resolution unanimously in 1997 that stated the USA would not agree to an international treaty on the reduction of GHG that does not involve developing countries and would harm the American economy. Subsequently, US President Clinton and Vice President Gore did sign the Kyoto treaty, but never forwarded it to the Senate (see Chap. 5).

A second dimension within the developed countries is political ideology and views on the environment. Harper (2012) differentiates between the view of a manifest destiny, which presumes a primacy of humans over the environment, and a view of environmental primacy. The former is more common among conservative parties, whereas the latter is more common among liberal and progressive parties. Hence, considering the timing of Kyoto ratifications, they are often aligned with the rule of a center-left party and a delay tactic or withdrawal during the rule of a conservative party. This was the case in Australia, where the Liberal Party under Howard refused to sign the Kyoto treaty, whereas after the change to the Labor Party, in 2008, Australia signed and ratified this treaty.

4.3 Non-governmental Organizations

After summarizing the most important developments in terms of environmental treaties and human rights issues at the international governmental level, this section now introduces a few non-governmental organizations in these areas. AI, HRW, and Oxfam will be used to describe the development in the human rights, which can be characterized by an increasing scope. As for environmental organizations, a distinction between local grassroots organizations and established environmental organizations is drawn, and subsequently the three organizations Sierra Club, Friends of Earth, and Greenpeace are discussed in more detail.

Human Rights Movements

Human rights movements cover a wide array of different topics and groups. Brysk (2013, 1) defines them as 'organizations, campaigns, and issue-networks that seek to enact the fundamental rights inscribed in the Universal Declaration of Human Rights.' This definition covers groups and activists that act at the local level, such as recent organizations dealing with disappearing persons, up to international organizations that contribute to legal frameworks such as the different UN human rights conventions.

Well-known international human rights organizations are AI and HRW, which were established with the goal of observing the human rights compliance of different governments (Amnesty 2017; Human Rights Watch 2017). More recently, human rights issues have been addressed by other organizations as well, such as Oxfam, which was founded as a development organization focusing on malnutrition, and the Sierra Club, which was traditionally an environmental organization (Brysk 2013). All of these organizations have been around for more than 50 years, and the Sierra Club since 1892. The shifting focus of human rights from civic and political rights to questions of discrimination due to age, disabilities, sexual orientation, and so on, are thus mirrored in the shifting foci of these organizations, as discussed in the following overview.

AI—which Brysk (2013) calls the 'flagship global organization'—was founded in London in 1961 by the lawyer Peter Benenson. AI consists mostly of volunteers, with groups of sufficient size being organized in sections, and overseen by the International Council. Currently, according to its webpage, 'Amnesty International is a global movement of more than 7 million people in over 150 countries and territories who campaign to end abuses of human rights' (Amnesty 2017). In sum, AI seems to have been quite successful, as it 'has secured the freedom of around one-third of its subjects from diverse regimes' (Brysk 2013).

AI's main activities are monitoring and reporting, with its annual reports on the state of human rights in different countries being well received and reported internationally. Furthermore, this INGO also encourages its members and all interested individuals to send letters, petitions, and emails to authorities, governments, and other political leaders. As of 19 April 2017, the AI webpage has calls such as sending a prewritten email to the US President Donald Trump to revoke his recent Executive Order, signing a petition for the victims of the conflict in Syria, or sending a premade message against the death penalty to all senators in the Philippines.

The trajectory of topics in human rights is reflected in the activities of AI. It was founded when Peter Benenson was outraged about the infringement of Art 18 of the Universal Declaration of Human Rights on the freedom of thought, conscience and religion and Art 19 on the freedom to change and manifest one's religion or beliefs. Opposition to authoritarian regimes and arresting of political opponents thus was the starting point. A look at the current list also mentions discrimination due to 'race, ethnicity, nationality, class, caste, religion, belief, sex, language, sexual orientation, gender identity, age, health or other status' and other issues (Amnesty 2017) and goes well beyond the topic of opposition to governments.

HRW was established in 1978 with the goal to check the compliance of the then socialist bloc with the Helsinki Accords and thus was initially called Helsinki Watch (Human Rights Watch 2017). The Helsinki Accords were an agreement between the West and the Communist bloc with the goal to reduce the tensions between these regions. It included an article on respect for human rights and fundamental freedoms, such as the freedom of thought, conscience, religion or belief, which was the focus of HRW. Over time, similar 'watches' were established for other areas of the world such as Latin America and Africa and all these different regional associations were combined as HRW in 1988.

Similar to AI, HRW extended its aims over time (see Human Rights Watch 2017). In the 1990s, topics such as domestic violence and human trafficking were included and most recently questions of health. HRW shares with AI a naming and shaming approach, which points to violations and publicizes them. HRW, in contrast to AI, relies more on experts and less so on individual memberships (Brysk 2013).

Such an extension of foci is also visible at Oxfam. This INGO was founded in 1942 in Oxford, UK, with a focus on a famine in Greece during World War II and continued to work on famine problems all over

Europe after the war (Oxfam 2017). Later on, Oxfam organizations were established in other countries and organized in the international cooperation Oxfam International in 1995. Oxfam International has currently 20 Oxfam affiliates from mostly developed countries plus two public engagement offices in South Korea and Sweden. It, however, reaches out beyond these geographic locations, as it provides support for local organizations in other countries. Oxfam now also deals with education, health, violent conflicts, migration, climate change, and other topics (Oxfam 2017). Just as the AI and HRW, Oxfam thus has expanded its scope considerably and goes well beyond its initial famine emphasis.

Environmental Movements

Environmental concerns go back to the nineteenth century when preservation and conservation groups became active, which now have been extended by groups with concerns about environmental justice and deep ecology views (Harper 2012; Rootes and Brulle 2013). Given this wide array of ideologies, the types of environmental organizations are also diverse and can include political parties, lobbying groups, direct action networks, and local neighborhood groups.

Local neighborhood groups are usually concerned with some immediate threat. The prime example in this regard is the Love Canal incident in Niagara Falls, New York (Harper 2012). Houses were built on top of an old chemical waste site and the chemicals leaked into the basements and surfaced on other places in this neighborhood, causing serious health problems. A local group, under the leadership of Lois Gibbs, became active and fought successfully against the authorities. Gibbs later founded the Citizens' Clearinghouse of hazardous waste, now the Center for Health, Environment, and Justice, and thus changed from a grassroots movement leader to a founder of a larger environmental organization.

Grassroots movements are typically reactions to specific local threats, consist of volunteers of immediately affected individuals, and cease to exist once the problem is solved (or cannot be solved). Contrary to such small groups, larger established environmental organizations usually focus on a wider range of topics. They, however, can be distinguished according to their approaches (Harper 2012), with reform groups focusing on lobbying and influencing political leaders and businesses, whereas direct action groups address the cause of the problem.

The following section introduces three organizations, of which Sierra Club can be considered the most reform-oriented group and Greenpeace the most direct-action-oriented group. Friends of the Earth International, on the other hand, is a loose network of groups with different approaches and is included as it started as a split off from Sierra Club. More extreme environmental groups such as Earth First! and the Earth Liberation Front that also call for violent actions against construction sites, machinery, and so on are not discussed. Yet, as Rootes and Brulle (2013) point out, activists in such groups still often share common identities with those of more reform-oriented groups.

Sierra Club, with a founding year of 1892, is one of the oldest environmental organizations in the USA (see Harper 2012; Sierra Club 2017). It is organized hierarchically in local and regional chapters with an overarching federal level. Despite its strong focus on reform approaches, the Sierra Club webpage also has a 'take action' section, which asks readers to take action on various issues. As of 18 April 2017, the call was 'Tell Your Senators to Defend the EPA,' which targets supposed plans of the Trump administration to eliminate the Environmental Protection Agency. Sierra Club thus combines its lobbying actions with organizing political actions, which also include frequent support of political candidates and parties via recommendations to Sierra Club members.

In 1969, after disputes between the organization and its executive David Brower, revolving around his opposition to a nuclear power plant near San Luis Obispo, Brower, supported by Robert Anderson, founded the organization Friends of the Earth (Saunders 2013). This organization turned into an international organization in 1971 and is now a federation of 75 different groups Friends of the Earth (International) (FOEI 2017). In contrast to other environmental organizations, the international network of FOE organizations is much more diverse, as membership has only minimal requirements (Saunders 2013). Consequently, the goals and targets of the different national organizations range widely. The common denominator is expressed in the main goals of FOEI (2017), which are climate justice, economic justice, food sovereignty, forests and biodiversity, and human rights, as topics.

Greenpeace was founded in 1971 in Vancouver, Canada, in conjunction with opposition to nuclear tests on an island near Alaska (Greenpeace 2017; Markham 2013). Since then, Greenpeace has extended its focus to climate change, marine life, sustainable farming, rainforests, and so on. It is an inherently international organization with a main office in Amsterdam and regional offices across the world, and supposedly 2.8 million supporters worldwide. In contrast to Friends of the Earth, the international organization and local offices are linked so tightly 'that their emphases and activities are difficult to distinguish' (Markham 2013).

Greenpeace describes itself online in its 'about us' section as 'an independent global campaigning organisation that acts to change attitudes and behaviour, to protect and conserve the environment and to promote peace' (Greenpeace 2017). Greenpeace is known for its media-focused campaigns that showcase environmental actions such as when a few core activists climb on ships, oil platforms, smoke stacks, waste sites, and other critical places. These actions are performed by professionals and then marketed to media and supporters (Markham 2013).

A very successful campaign in terms of altering behavior was the 1995 campaign against the oil storage facility Brent Spar, which Shell wanted to sink in the Atlantic Ocean (Greenpeace 2017). Greenpeace warned that it would cause damage to the ocean and called for a disposal on land, including dismantling the buoy. The INGO managed to organize a consumer boycott of Shell gas stations in Germany and other northern European countries and also to gain the support of various politicians. Shell gave in and transported the buoy to Norway for dismantling. Later it turned out that the remaining oil quantity was overestimated and the environmental risks much smaller than anticipated. Yet it is a prime example of the successful mobilization of people.

4.4 Summary

This chapter provided an overview of the international level with an emphasis on the UN, related treaties and covenants, and INGOs in the human rights and environment arenas. World Society scholars assume that this international level influences outcomes at a national and regional level that reflect the models and ideas emphasized by the institutions. In this vein, several authors have studied their effects, usually relying on a time-comparative approach.

As for human rights, Koo and Ramirez (2009) showed that domestic human rights institutions are implemented more often, among other factors, when ties to World Society are strong, and in time periods before and after international conferences. Cole (2012a) found that the International Covenant on Civil and Political Rights has had positive effects on basic civil rights, whereas physical integrity abuses did not change much. Cole (2012b) also looked at the effects of the Convention on the Elimination of Discrimination against Women and concluded that it had positive effects on political rights, whereas economic rights were unaffected, and the impact on social rights even negative.

As for environmental institutions, Shorette (2012) found that links to World Society affect the use of chemical fertilizers and pesticides, with different effects depending on geopolitical position of a country in the core, semi-periphery, or periphery. Hironaka (2003) found positive effects of IGO membership on the adoption of environmental impact assessment legislation, and Shandra et al. (2012) found that environmental INGOs reduce the number of threatened mammals in less developed countries.

Frank et al. (2000) considered a wide range of possible impacts of global forces and used the number of national parks, numbers of chapters of international environmental NGOs, the number of environmental IGOs a country belongs to, the adoption of environmental impact assessment legislation, and the founding of environmental ministries as dependent variables. All these variables were positively influenced by ties to World Society in terms of IGOs and INGOs, number of receptor sites in terms of number of domestic ecology associations and natural science associations, as well as the size of global institutionalization in terms of staff size in UNEP and the occurrence of the 1972 and 1992 UN environment conferences.

In sum, these publications show that the international level has manifold influences on domestic outcomes that can range from changes in pesticide use to the founding of environmental ministries on environmental issues, and from improvement in the social lives of the population to the establishing of human rights institutions. Yet, the implementation of global ideas can also be influenced by the national political system, economic circumstances, and other domestic factors. The next chapter, thus, looks into the specific circumstances of the USA, Canada, Germany, Austria, Czech Republic, and Russia.

Notes

1. Further examples are the International Convention on the Suppression and Punishment of the Crime of Apartheid (1973), Convention on the Rights of the Child (1989), Convention Against Torture (1984), Convention on the Elimination of All Forms of Racial Discrimination (1969), Convention on the Elimination of All Forms of Discrimination Against Women (1979), International Convention on the Protection of the Rights of All Migrant Workers and Members of Their Families (2003), Convention on the Prevention and Punishment of the Crime of Genocide (1948), Convention on the Rights of Persons with Disabilities (2007), International Convention for the Protection of All Persons from Enforced Disappearance, 2006 Indigenous and Tribal Peoples Convention, 1989 (ILO 169). Source: OHCHR (2017).
2. https://www.theguardian.com/world/2004/oct/23/society.russia (access 12 April 2017).

References and Further Readings

Amnesty. 2017. https://www.amnesty.org/. Accessed 19 Apr 2017.

Appadurai, A. 1996. *Modernity at Large: Cultural Dimensions of Globalization*. Minneapolis: University of Minnesota Press.

Brysk, A. 2013. Human Rights Movements. In *The Wiley-Blackwell Encyclopedia of Social and Political Movements* (Online).

Burke, R. 2008. From Individual Rights to National Development: The First UN International Conference on Human Rights, Tehran, 1968. *Journal of World History* 19 (3), *New Histories of the United Nations* (Sep 2008), 275–296.

Cairo Declaration. 1990. http://hrlibrary.umn.edu/instree/cairodeclaration.html. Accessed 21 Apr 2017.

Canan, P., S. Andersen, N. Reichman, and B. Gareau. 2015. Introduction to the Special Issue on Ozone Layer Protection: The Extraordinary Experience of Building the Montreal Protocol, Lessons Learned, and Hopes for Future Climate Change Efforts. *Journal of Environmental Studies and Sciences* 5 (2): 111–121.

Cole, W.M. 2012a. Institutionalizing Shame: The Effect of Human Rights Committee Rulings on Abuse, 1981–2007. *Social Science Research* 41: 539–554.

———. 2012b. Government Respect for Gendered Rights: The Effects of the Convention on the Elimination of Discrimination Against Women on Women's Rights Outcome, 1981–2004. *International Studies Quarterly*: 1–17.

FOEI. 2017. http://www.foei.org/about-foei/history. Accessed 17 Apr 2017.

Frank, D.J., A. Hironaka, and E. Schofer. 2000. The Nation-State and the Natural Environment over the Twentieth Century. *American Sociological Review* 65 (1): 96–116.

Gareau, B.J. 2013. *From Precaution to Profit: Contemporary Challenges to Environmental Protection in the Montreal Protocol*. New Haven, CT and London: Yale University Press.

Greenpeace. 2017. http://www.greenpeace.org/. Accessed 17 Apr 2017.

Grundmann, R. 2001. *Transnational Environmental Policy: Reconstructing Ozone*. London: Routledge.

Hanhimäki, J.M. 2015. *The United Nations: A Very Short Introduction*. Oxford: Oxford University Press.

Hardin, G. 1968. The Tragedy of the Commons. *Science, AAAS* 162 (3859): 1243–1248.

Harper, Charles. 2012. *Environment and Society. Human Perspectives on Environmental Issues*. 5th ed. Boston, MA: Pearson.

Hironaka, A. 2003. Science and the Environment. In *Science in the Modern Polity. Institutionalization and Globalization*, ed. G.S. Drori, J.W. Meyer, F.O. Ramirez, and E. Schofer, 249–279. Stanford, CA: Stanford University Press.

———. 2014. *Greening the Globe. World Society and Environmental Change*. Cambridge: Cambridge University Press.

Human Rights Watch. 2017. https://www.hrw.org/history. Accessed 19 Apr 2017.

Jorgenson, A., C. Dick, and J. Shandra. 2011. World Economy, World Society, and Environmental Harms in Less-Developed Countries. *Sociological Inquiries* 81: 53–87.

Kennedy, P. 2006. *The Parliament of Man. The United Nations and the Quest for World Government*. London: Allen Lane.
Koo, J.-W., and F. Ramirez. 2009. National Incorporation of Global Human Rights: Worldwide Expansion of National Human Rights Institutions 1966–2004. *Social Forces* 87 (3): 1321–1353.
Kyoto Protocol to the United Nations Framework Convention on Climate Change. 1998. https://unfccc.int/kyoto_protocol/items/2830.php. Accessed 10 Apr 2017.
Markham, W. 2013. Greenpeace. In *The Wiley-Blackwell Encyclopedia of Social and Political Movements* (Online).
Meisler, S. 1995. *United Nations. A History*. New York: Grove Press.
OHCHR. 2017. http://www.ohchr.org/EN/ProfessionalInterest/Pages/CoreInstruments.aspx. Accessed 11 Apr 2017.
Oxfam. 2017. https://www.oxfam.org/en/about. Accessed 27 Apr 2017.
Paris Agreement. 2015. United Nations. http://unfccc.int/files/essential_background/convention/application/pdf/english_paris_agreement.pdf
Rootes, C., and R. Brulle. 2013. Environmental Movements. In *The Wiley-Blackwell Encyclopedia of Social and Political Movements* (Online).
Saunders, C. 2013. Friends of the Earth. In *The Wiley-Blackwell Encyclopedia of Social and Political Movements* (Online).
Shandra, J., L. Esparza, and B. London. 2012. Nongovernmental Organizations, Democracy, and Deforestation: A Cross-National Analysis. *Society and Natural Resources* 25: 251–269.
Shandra, J., C. Leckband, L. McKinney, and B. London. 2009. Ecologically Unequal Exchange, World Polity, and Biodiversity Loss: A Cross-National Analysis of Threatened Mammals. *International Journal of Comparative Sociology* 50: 285–310.
Shorette, K. 2012. Outcomes of Global Environmentalism: Longitudinal and Cross-National Trends in Chemical Fertilizer and Pesticide Use. *Social Forces* 91: 299–325.
Sierra Club. 2017. http://www.sierraclub.org/about. Accessed 18 Apr 2017.
Stiglitz, J.E. 2002. *Globalization and Its Discontents*. New York: W.W. Norton.
Taraska, G. 2015, Dec 15. *The Paris Climate Agreement* (PDF). Center for American Progress. https://cdn.americanprogress.org/wp-content/uploads/2015/12/15030725/ParisClimateAgreement.pdf. Accessed 10 Apr 2017.
UN Durban Review. 2017. http://www.un.org/en/durbanreview2009/. Accessed 11 Apr 2017.

UN Treaty Collection. 2016. https://treaties.un.org/Pages/ViewDetails.aspx?src=TREATY&mtdsg_no=IV-2&chapter=4&clang=_en. Accessed 29 Nov 2016.

UNFCCC. 2017. https://unfccc.int/2860.php. Accessed 11 Apr 2017.

Union of International Associations. 2017. http://www.uia.org/faq. Accessed 11 Apr 2017.

Wallerstein, I. 2004. *World-Systems Analysis: An Introduction*. Durham, NC: Duke University Press.

Weart, S. Dec 2011. *International Cooperation: Democracy and Policy Advice (1980s). The Discovery of Global Warming.* American Institute of Physics. Accessed 7 Apr 2017.

5

Development in Selected Countries

The previous chapter offered an overview of developments at the international level with a view on human rights and environment-related institutions within the United Nations, international treaties such as the Kyoto Protocol and the Paris Agreement, and also on some INGOs in detail. The current chapter now looks at a few selected countries and discusses the specific national developments as well as the individual attitudes and behaviors in the USA, Canada, Germany, Austria, Russia, and the Czech Republic.[1] This country selection follows the classification of polities with regard to the role of the individual in the policy process, proposed by Schofer and Fourcade-Gourinchas (2001).

The USA is selected to represent liberal regimes with strong demands on individual participation and few environmental regulations. Germany represents corporatist regimes where the state takes over many individual responsibilities and imposes strict laws. Russia represents former socialist countries with corporatist regimes that have experienced rapid transitions over the last 25–30 years. In addition, a similar, yet slightly different case is added to each group—Canada to the USA, Austria to Germany, and the Czech Republic to Russia.

The description of environmental policies and changes over time will point to the influence of political parties and incumbents. When

M. Hadler, *The Influence of Global Ideas on Environmentalism and Human Rights*,
DOI 10.1057/978-1-137-57440-4_5

discussing their different environmental positions, it is worth considering the underlying environmental beliefs and values. Harper (2012) differentiates between two views, believing in a manifest destiny and believing in an inherent value in the environment. Manifest destiny refers to the position that humans are supposed to use the environment for egocentric reasons, whereas the latter refers to the opposite. These views are reflected in various degrees in the positions of political parties, with conservative parties leaning to the former, while particularly Green parties lean toward the latter.

The empirical results in terms of environmental attitudes and behaviors presented in this chapter are based on the International Social Survey Programme (ISSP) data from 1993, 2000, and 2010. The overview of the country developments thus focuses on the period of 1993 onward, but also makes references to previous events when necessary. The first section starts out with an overview of societal developments and the perception of different environmental threats in these selected countries. These overviews are followed by more detailed accounts of national developments, which, then, are related to the changes in environmental behaviors over time in the concluding section.

5.1 Differences in Environmentalism and Related Factors at the Country Level

The main theme of this book is environmentalism and the influence of global cultural models on related behaviors. Therefore, the overview of country trends presented in Table 5.1 starts with the magnitude of public and private environmental behaviors, their changes over time, and ties to world society. These items are followed by indicators of societal affluence, pollution, and environmental impact—measures that are related to theories such as ecological modernization, value change, the difference between subjective environmental values and objective environmental threats.

Public behavior is based on the four ISSP items: being a member of an environmental organization, having taken part in a protest, donating money to an environmental group, and signing a petition for environ-

Table 5.1 Changes over time in environmental behaviors and context characteristics in six countries

		Public behavior	Private behavior	IGOs	INGOs	GDP per capita	PM10 exposure	CO_2 output per head	Ecological footprint
USA	1993	34%	2.21	64	2226	25,361	29.61	19.9	7.8
	2010	26%	2.35	63	3203	46,546	18.00	17.3	8
	% change	−22%	6%	−2%	44%	84%	−39%	−13%	3%
Canada	1993	57%	2.35	65	1981	19,637	24.78	16.6	5.3
	2010	42%	2.96	62	2749	46,361	15.00	15.2	7
	% change	−26%	26%	−5%	39%	136%	−39%	−8%	31%
Austria	1993	48%	3.07	60	2174	24,016	38.66	7.6	4.5
	2010	35%	3.03	66	3171	45,158	27.00	7.4	5.3
	% change	−26%	−1%	10%	46%	88%	−30%	−3%	17%
Germany (West/East)	1993	39%/34%	2.91/2.82	81	2911	24,805	26.81	10.8	4.99
	2010	31%/22%	3.09/3.04	86	4215	39,857	16.00	9	5.1
	% change	−19%/−34%	6%/8%	6%	45%	61%	−40%	−17%	2%
Czech Republic	1993	22%	2.08	42	1359	3766	41.32	12.8	4.9
	2010	19%	2.61	60	2484	18,839	16.00	10.3	5.7
	% change	−14%	26%	43%	83%	400%	−61%	−20%	17%
Russia	1993	19%	1.50	39	514	2779	40.80	13.2	5.2
	2010	7%	1.59	71	2271	10,351	27.00	11.1	4.4
	% change	−62%	6%	82%	342%	273%	−34%	−16%	−15%

Sources: Behaviors (ISSP 1993 and 2010), International Organizations (UIA various years), GDP per capita in US$1000 (UN 2015), CO_2 output per capita and air pollution as particulate matter exposure of the population (World Bank 2015), and ecological footprint (UNEP 2014)

mental issue. The table displays the percentage of respondents who were active in at least one of these activities. Private behavior reflects the mean value based on an index considering the items recycling, and cutting back the use of a car, with 5 indicating frequent behaviors and 1 none.

As for the country-level characteristics, the number of ties to world society is measured by the number of international governmental organizations (IGOs) and international non-governmental organizations (INGOs); the level of societal affluence by GDP per capita; environmental pollution by the particulate matter exposure (PM10) of the population and the CO_2 output per head; and environmental impact by the 'ecological footprint' indicator. All values are presented for 1993 and 2010, and in addition, the relative change from 1993 to 2010 is reported.

Table 5.1 shows that in 2010 public behavior was the highest in Canada, followed by West Germany and Austria, whereas it is least frequent in Russia. Russians also show the least private behavior. The most prevalent is private behavior in Austria, followed by West and East Germany. In terms of changes over time, public behavior has declined in most countries, whereas private behavior has increased in most countries.

As for ties to world society, Western countries were better linked in 1993 than the Czech Republic and Russia, but over time, both former socialist countries caught up with the other countries—especially as far as IGOs are concerned. Similar developments occurred in societal affluence, where a large gap in 1993 was followed by a huge increase in affluence in the Czech Republic and Russia. The environmental pollution indicators show less variation across countries and similar improvements in most countries, with a 30–40% decline in PM10 exposure and a smaller decline in CO_2 output. The ecological impact, on the other hand, has decreased only in Russia—probably due to the decline in industry—whereas it increased in all other countries.

As for the environmental pollution indicators, we need to be aware that the CO_2 output is based on the production aspect and can be much larger when considering the consumption aspect. The latter also considers the CO_2 output that is outsourced to countries where the goods are produced, the transport of these goods, and other factors (Peters et al. 2010). As for Austria, for example, Munoz and Steiniger (2010) have estimated that the consumption-based indicator would be 44% larger than the production-based factor.

Before discussing the development in these countries, Table 5.2 offers an overview of perceived environmental threats, which allows a comparison between actual threats and the individual perceptions of threats. The percentages are based on the responses to the following statement: 'Here is a list of some different environmental problems. Which problem, if any, affects you and your family the most? (1) Air pollution, (2) Chemicals and pesticides, (3) Water shortage, (4) Water pollution, (5) Nuclear waste, (6) Domestic waste disposal, (7) Climate change, (8) Genetically

Table 5.2 Perceived environmental threats

	USA	Canada	Austria	Germany	Czech Republic	Russia	All 36 ISSP countries
Air pollution	24.5%	17.5%	16.4%	15.4%	29.4%	18.0%	19.5%
Chemicals and pesticides	13.0%	17.6%	7.7%	9.1%	8.8%	7.0%	10.9%
Water shortage	5.8%	2.6%	1.4%	1.6%	2.9%	5.2%	7.7%
Water pollution	10.1%	11.1%	6.5%	3.1%	5.7%	20.3%	11.4%
Nuclear waste	1.2%	1.6%	2.7%	6.5%	3.3%	6.3%	3.3%
Domestic waste disposal	5.3%	10.9%	12.1%	5.1%	14.3%	10.3%	11.5%
Climate change	6.9%	12.7%	20.7%	20.3%	10.1%	10.9%	11.5%
Genetically modified foods	10.8%	9.3%	15.3%	13.3%	3.7%	11.4%	9.0%
Using up our natural resources	12.3%	9.1%	5.6%	10.8%	5.8%	8.0%	7.2%
None of these	10.0%	7.5%	11.6%	14.7%	16.0%	2.6%	8.0%
N	1224	910	967	1245	1324	1532	46,659

Respondents were asked to choose a single item based on the question: 'Here is a list of some different environmental problems. Which problem, if any, affects you and your family the most?' Source: ISSP 2010 modified foods, (9) Using up our natural resources, (10) None of these.' Respondents had to choose a single item. So, in case there is a particular pressing issue present—such as a water shortage, for example—other issues remain ignored.

These questions were asked in a total of 36 countries, including less-developed countries such as the Philippines (see Hadler and Kraemer 2016, for a full overview). Among the total country sample, air pollution is mentioned the most often (19.5%), followed by climate change (11.5%) and domestic waste disposal (11.5%), chemicals and pesticides (10.9%), genetically modified foods (9.5%), water shortage (7.7%), using up our natural resources (7.2%), and nuclear waste (3.3%). Another 8% of the respondents stated that none of these problems is affecting them or their family.

Considering our six selected countries, when compared to the total average, respondents in the USA perceive air pollution and the overuse of natural resources as relatively strong threats. Canadians are relatively more worried about chemicals in the food supply and the use of natural resources. Austrians and Germans are particularly concerned about climate change and Genetically modified organism (GMOs), plus natural resources in Germany. Czechs mention air pollution and waste disposal, whereas Russians are worried about water pollution.

Analyses of this distribution across all 32 countries identified three underlying dimensions (Hadler and Kraemer 2016). The first one can be described as 'water issues versus all other problems.' The second one distinguishes between respondents who said that none of threats affect them and respondents who mentioned a threat. The third one discriminates between high risks such as nuclear waste, genetically modified food, using up of resources and climate change, and everyday problems such as air pollution, chemicals and pesticides, and domestic waste disposal.

Relating these dimensions to country-level characteristics showed that water issues were chosen less often in wealthier societies and EU member countries. High risks such as climate change, nuclear power etc. were mentioned more often in EU member countries than in non-EU countries with a similar level of affluence. This difference suggests that common environmental and other regulations may also have an impact on the distribution of environmental threats and their perception, beyond wealth and modernization effects.

5.2 Specific Trends in Selected Countries

Now that we have had a general overview of trends in our selected countries, we offer a more detailed summary of specific developments in these countries. The presentation follows the division in liberal, corporate welfare, and post-communist countries. For each country, the main environmental circumstances are summarized, the political system and positions of the ruling government, as well as the positions regarding international treaties. Furthermore, the political significance of Green Parties will be addressed, as they are institutionalized environmental movements (Rootes 2003), important political allies for any activists, and related to strong activism and concerns (Hadler and Haller 2011).

The USA

A very diverse environment ranging from coastal areas, large plains, deserts to mountainous regions characterizes the USA. Due to its size, it also has ample natural resources, such as oil, gas, forestry, fish, and so on. The USA is known for its enormous energy consumption and the related output of greenhouse gases (GHG). Considering the UNEP (2015) report on such emissions, the USA is the second highest emitter after China among the G20 countries, with the total output having increased from 1990 to 2010. The per capita output, on the other hand, has declined, which shows the strong influence of a growing population on the absolute output.

Environmental protection has a long history in the USA, with preservation and conservation aspects as the dominant thoughts in the nineteenth century (Harper 2012). As in many other industrialized countries, problems such as air and water pollution became important topics in the 1960s and 1970s, with incidents such as the love canal's waste sites underneath houses, and other events leading to an increased awareness. This evoked two forms of environmental movements—reform environmentalism that focused on influencing political stakeholders to make changes and grassroot activism by neighborhood groups that were immediately affected.

The salience and public awareness of these topics in the early 1970s are reflected in the fact that the Republican President Nixon began by announcing that he was an environmentalist (Harper 2012, 231). Nixon also founded the federal US Environmental Protection Agency (EPA), which could build upon earlier environmental laws and protection measures (Andrews 2006). By contrast, the emphasis on environmental issues is currently low. Scott Pruitt, who in 2017 was selected by the Republican President, Donald Trump, as the head of the EPA is described by the *New York Times* as 'a key architect of the legal battle against Mr. Obama's climate change policies, actions that fit with the president-elect's comments during the campaign.'[2]

Two parties dominate the US political system: Republicans and Democrats. The Republican and Democratic parties differ in their view on environmental issues, with the former showing stronger support for the paradigm of 'manifest destiny' which calls for the use of natural resources for humankind (Harper 2012), while the latter is somewhat less supportive of this view. The period under investigation (1993–2010) began with the Democratic President Bill Clinton, who succeeded the presidencies of Ronald Reagan (1981–1989) and George H. W. Bush (1989–1993). Clinton had a stronger environmental commitment than did the previous presidents and also his successor George W. Bush (2001–2009). The Obama years (2009–2017) were again more environmentally focused, whereas the current president, Donald Trump (2017–), does not appear to be a forerunner in environmental issues.

Given the dominance of the Democrats and the Republicans, smaller parties such as the Green Party and independent candidates do not have much influence at the federal or state level. Strong candidates instead make negative news and are seen as taking away votes from the Democratic Party, as in the 2000 presidential election, in which Ralph Nader received almost 3% of the popular vote, which allegedly cost Al Gore the presidency. Yet, the Green Party has had some success at the local level, particularly in the Pacific Coast states, and has in total more than 300,000 registered members.

At the international level, the USA focuses on the impact of environmental treaties on the domestic economy. Bill Clinton and his Vice President Al Gore signed the Kyoto Protocol, but did not submit it to the Senate, because it was clear to Clinton and Gore that the resolution would never pass Senate (Hovi et al. 2010). His successor, George W. Bush, withdrew from the treaty because of possible negative effects for the US economy. Bush's environmental policy remained weak and only under Obama in 2009 did environmental topics gain more momentum, which is seen in the fact that the USA and China declared jointly in 2016 that they will sign the Paris Agreement.[3]

Canada

Canada is also a large country with ample natural sources of coal, oil, and wood. Like the USA, it is a major emitter of GHG, one of the top per head GHG emitters among the G20 countries and in 2010 emitted more per head than the USA (UNEP 2015). In terms of absolute emissions, however, Canada is low. Then again, the export of coal also affects the output of other countries, such as China, which is the top emitter among the G20 in terms of total output, as pointed out within the literature on the difference between production- and consumption-based calculations of the CO_2 output (Peters et al. 2010).

In the past, Canada faced similar problems as in other developed countries, with acid rain, toxic waste, air and water pollution, and other typical problems of the industrialized world (see Hadler and Kraemer 2016). As in other countries, environmentalism increased in public awareness in the 1960s and 1970s around topics of nuclear power, acid rain, and waste, which led to the foundation of Greenpeace in Vancouver in 1971. Other than that, Canada underwent typical phases of environmentalism and associated actions, with, for example, the founding of the Department of Environment (now Environment and Climate Change Canada) in 1971 by the Liberal Party government (Olive 2015).

The federal structure of Canada leaves considerable power in the hands of the states and provinces, which can hamper the work of the federal environmental ministry. Van Nijnatten (1999) concludes in her comparison of the USA and Canada that the aggressively independent provincial governments are a main obstacle in environmental protection measures and efforts, as the provinces have jurisdiction in energy matters. On the other hand, some provinces can implement their own programs, such as seen in British Columbia, where the Liberal governor proposed and successfully implemented a carbon tax (Harrison 2012). Federal and provincial politics thus can both foster and hamper environmental efforts.

Examining the federal government, the period of investigation starts with the end of the Progressive Conservative federal government, when the Liberals won the election in 1993 and remained in power until 2006. In 2006, the Conservatives were able to win and governed until the Liberals won again in 2015. Canada's environmental positions shift with the ruling party, and are more pro-environment during liberal governments and less pro during conservative rule, with the exception of the pre-1993 period when the Progressive Conservatives were in power. This party favored environmental issues, which can be seen in the fact that their then environmental advisor, Elizabeth May, is now leader of the Green Party.

Alongside these main parties, the Green Party is also active. It was established in 1983 and ever since faces the problems of a majority system. In the elections of the representative house, the Green Party had a peak in 2008 when they reached around 7% of the public vote. This 2008 election had a strong focus on environmental topics, with the Liberals proposing a 'Green Shift,' given that environmental issues gained a strong public awareness around 2006 when first effects of climate change were noticed (Harrison 2012). The Green Party lost votes in the following elections. Yet, the leader of the Green Party, Elizabeth May, was able to win a seat in Vancouver Island in 2011 and thus get a seat in the lower house—despite a nationwide low of 3.5% of the vote in the 2015 election.

As for the main international treaty during the period under investigation, the Kyoto Protocol, Canada was initially a main proponent, with the Liberal government signing the treaty in 1997 and ratifying it in the parliament in 2002. Yet, the government showed little signs of willingness to implement changes, and with the change to the conservative government in 2006, Canada became more wary of this treaty (Harrison 2012). It withdrew its participation in the Kyoto Protocol, which was announced in Dec 2011 by the Minister of the Environment, Peter Kent, Conservative, with the argument of it being ineffective, as the two main emitters, USA and China, are not covered. Yet, Canada signed the Paris Agreement and pledged to lower its greenhouse gas emissions by 30% by 2030 compared to the 2005 level.[4]

Germany

The central European country Germany is much smaller than the USA and Canada and thus environmentally less diverse. Natural resources in fossil fuels such as oil, gas, and coal have lost importance over the last decades and have now much less significance than in the USA and Canada. Energy consumption is lower as well and so is the output of GHG, as can be seen in Table 5.1.

Germany is considered a corporative state, one which emphasizes the role of the state and less so the role of the individual (Schofer and Fourcade-Gourinchas 2001). This results in tighter environmental regulations. Germany, however, has a very specific political history, as it was split after World War II into the Federal Republic of Germany (West Germany) and the German Democratic Republic (East Germany). Both countries were reunited in 1990, yet the legacy of the socialist past is still visible in the Eastern part. Even now—consider Table 5.1—the data indicate that environmental behaviors are still lower in the former socialist part.

Both West and East Germany face the typical environmental problems of industrialized countries. Compared to the USA, nuclear power has been of greater importance due to the major nuclear accident in Chernobyl in 1986, which caused radioactive fallout across Europe and thus was much more widespread than the five-mile incidence in the USA. Furthermore, it is important to mention that the end of the socialist system in Eastern Germany also resulted in a major recession and a decline in industry and associated emissions in this part of Germany, which allowed Germany to make strong commitments in emissions reductions in international treaties.

The western and the eastern parts of Germany also have different trajectories in terms of environmental movements (see Jones 1993; Markham 2008). In contrast to other former socialist countries, the environmental movement was active during the last decade of East Germany, where the environmental movement worked together with the human rights movement and the church. In West Germany, the environmental movements were able to confront the ruling government much more directly. The early confrontations in the 1980s revolved around nuclear power plants, airport extensions, and the Rhine-Danube channel. The movement became institutionalized in West Germany soon after: Greenpeace Germany was founded in 1980, an environmental ministry was established in 1986, and the Green Party won seats in the national parliament, the 'Bundestag,' for the first time in 1983, whereas in East Germany the Green Party did not become part of the political system until 1990.

The focus of this book is on the period after 1990, thus only the political system of Germany (the former West Germany) is described. Its political system consists of smaller parties and therefore better opportunities for becoming established than the party systems of the USA and Canada, as some of the votes are distributed proportionally based on the results of a general election. Furthermore, given that the ruling party needs the support of the parliament and parties are rarely able to gain a majority, all federal governments since World War II have been coalitions of different parties. This also resulted in the participation of the Green Party from 1998 to 2005, when it was in a coalition with the Social Democrats. This success is even more interesting, considering that after the reunification of Germany, the Green Party remained in the federal parliament only due to the eastern part of Germany since it failed to achieve the 5% hurdle in the western part in the 1990 elections.

Internationally, Germany plays in important role in environmental issues. It hosts the secretariat of the United Nations Framework Convention on Climate Change (UNFCC), organized the first United Nations Climate Change conference in 1995, and signed various international treaties. Within Kyoto, Germany pledged to reduce its emission by

21%, whereas the EU 15 pledged to reduce it by 8%. However, as mentioned before, Germany had the advantage that the economy in the eastern part had collapsed and the emissions were already much lower than before. More recently, Germany also signed the Paris Agreement, which passed the German Bundestag in a unanimous vote.

Austria

Austria is about a quarter of the size of Germany and a predominantly mountainous country. Its economy was historically focused on raw materials and experienced an economic boom after World War II that was characterized by little emphasis on environmental protection. Over time, problems with pollution became more visible and air and water quality issues prevalent. Two specific events are often mentioned as crucial for the increase in environmental awareness: a referendum in 1979, in which the public decided—against the government's preference—to not bring the finished nuclear power plant in Zwentendorf into operation; and the 1984 protest against the construction of a hydropower plant near Vienna (see Haller and Troy 2003; Pesendorfer 2007).

While such public protests do happen, Austrian society and policy is characterized by a strong emphasis on the role of the state and the government. Reform-oriented environmental movements, which try to lobby and influence governments, thus play an important role. This relationship was visible when comparing the development of the opposition to GMOs in Austria and France (Seifert 2002). In Austria, a compromise between NGOs and the governments was established and GMOs were banned, whereas in France radical activism, including burning down of fields, occurred.

As for the regular political system, Austria shares many similarities with Germany. Seats in the parliament are distributed proportionally, based on the results of a general election. The party with the most seats is invited by the president to form a government, but given that the government depends on a majority in the parliament, coalitions are common, with the last single party government that held a majority in the parliament ruling from 1971 to 1983.

Most governments since then were coalitions between center-left and center-right parties. The positions of the different governments and the importance of the environment thus has changed less than in the two party systems of the USA and Canada. Yet, the different governments and economic circumstances have had influences on environmental politics. The Ministry of Environment, for example, was founded in 1972 and initially held by the Social Democrats. In this period, this ministry also focused on health. Later, when the environment ministry was moved to the conservative party, health was moved to a different ministry and the environment agenda combined with youth and family, and later with agriculture and forestry.

Overall, the political system and electoral rule of Austria offer good opportunities for smaller parties such as the Green Party. The Green Party has been part of the national parliament since 1986 when it reached 4.8% of the votes. In 2013, the last general election, the Green Party was able to secure 12.4% of the votes. In contrast to the German Green Party, the Austrian party has never been part of the federal government. Yet, it has had more success at the provincial and local government level where the Greens have been part of many different governments.

The influence of Austria on international politics is limited given the size of country. Nonetheless, Austria is home to the UN headquarters of International Atomic Energie Agency (IAEA), United Nations Industrial Development Organization (UNIDO), and Preparatory Commission for the Comprehensive Nuclear-Test-Ban Treaty Organization (CTBTO) and has hosted various international conferences such as the human rights conference in 1992 and has signed the major environmental treaties. The Kyoto Protocol, for example, was signed in 1998 and ratified in 2002. Austria had pledged to reduce its emissions by 13%, but was able to achieve this only by buying certificates for 500 million euros in order to cope with the excess of 70 million tons of CO_2 emissions.[5] The Paris Agreement was finally passed by parliament with the consent of all parties except for the rightwing Freedom party.[6]

The Czech Republic

The Czech Republic is a neighboring country of Austria and Germany and shares many geographical characteristics with these countries.

Differences in environmental problems thus rather resulted from its socialist past. The Czech Republic—as part of Czechoslovakia until the separation of the Czech Republic and Slovakia in 1993—belonged to Eastern European socialist countries after World War II, economies which gave environmental protection little emphasis.

The socialist economy relied heavily on the use of brown coal, which resulted in severe problems with acid rain and SO_2 emissions. Furthermore, contamination of soil due to poor treatment of hazardous waste was a problem. Now, more than 25 years after the transition, environmental problems have improved, but air quality is still a major issue. This is seen objectively in the State of the Environment Report (EEA 2017) and also among the perceived risks of the population (see Table 5.2).

As indicated above, the improvements in environmental pollution were initially caused by the transition to a market-based economy, which resulted in a decline in industrial production and the associated pollution despite little political emphasis on environmental issues (Schreurs and Papadakis 2007a). Later on, however, environmental problems were also addressed by the political system because of the accession process to the European Union (EU), which involved the implementation of EU law and environmental standards.

Apart from the negative impact on the environment, the socialist system also had negative effects on environmental movements. The activities of these movements, however, accelerated enormously during and after the breakdown of the socialist system (see Fagan 2004). An environmental ministry was established in 1989 and environmental groups—just like in Russia—received support from the West during the transition period. This emphasis, however, was lost during the Klaus government (1992–1997), in which 'the environmental movement was politically marginalized and castigated as an anti-market relic of the socialist era' (Fagan 2004, 5).

The trajectory of the Czech Green Party followed these developments (see Schreurs and Papadakis 2007b). It was founded in 1990, had some immediate success, and won three seats in the parliament. Soon after, the political focus shifted to development issues and the separation from the Slovak Republic, which resulted in rather bleak years for the Green Party in the 1990s. Later, the Green Party was more successful and was able to win a Senate seat in the Upper House (2004), had governmental participation from 2007 to 2009, and is currently holding three seats in the Senate.

Internationally, the breakup of the Eastern European socialist bloc and the separation from the Slovak Republic were decisive events for the role of the Czech Republic at the international level. As for the environment, the Kyoto Protocol was signed in 1998 and ratified in 2001. The Czech Republic did not have problems meeting the Kyoto target, due to the collapse of the socialist industry and the aforementioned improvements. Yet, considering changes in the ecological footprint (Table 5.1) the Czech Republic has increased more than in Russia as the social and economic transition went rather well. The Paris Agreement was signed by the Czech Environment Minister on 22 April 2016, yet the agreement is not ratified and thus not in force yet.[7]

Russia

Russia and former Soviet countries' environmental problems were widespread and related to a low emphasis on environmental protection in the government. Issues ranged from industrial pollution, deforestation, to nuclear waste, but were also caused by large building projects such as the attempt to reverse Siberian rivers and the threat of Lake Baikal (Yanitsky 2012). Environmental protest against such projects and their negative environmental effects was one of the first forms of political protest against the communist system, with a democratic upsurge during the regime transition in the late 1980s and early 1990s.

The transitions of the last 30 years also caused major changes to the economic structure, social structure, and the amount of pollution. Accordingly, Table 5.1 shows that both pollution measures and the ecological footprint declined in Russia. As for the transition of the political system, Stykow (2014) points out that Russia's strong emphasis on the state and a concentration of power among small circles of politicians did not change. Russia thus cannot be seen as a pluralistic democratic regime, but rather as a system of strong single persons such as Boris Yeltsin and now Vladimir Putin.

Yet, Russia also has an environmental ministry. It was established in 1996 as a successor to the State Mining and Geological Service, which was established under Peter the Great in 1700. Other than that, the political circumstances leave only little room for environmental forces in the regular political system. The Green Party was founded in 1992 and has participated in various legislative elections. It, however, never received more than 1.4% of the votes and was banned from the 2007 elections due to alleged manipulations. In 2016, it reached only 0.73% of the votes in the Duma election.

As for environmental protest, Henry (2002) distinguishes two types of protests, grassroots movements and reform environmentalists. The latter, which he refers to as professionalized environmental organizations, received their financial support from abroad, with about 75% of the means being transferred from the West. Henry suggests that the donations from the West also had a goal to influence on Russian politics—a goal, he concludes, that was not achieved, as these organizations are operating in a vacuum.

Henry's view is opposed to the WS view of organizations influencing national politics. Yet, his assessment was made more than 15 years ago. The recent legal changes in Russia, which require all not-for-profit organizations that receive money from foreign sources to register and to declare themselves as foreign agents (the foreign agent law of 2012), suggest some concerns about the influence of such organizations and may indicate that Henry's assessment of 2002 was premature.

Internationally, Russia ratified the Kyoto Protocol in 2004, which brought the treaty into effect, as it needed 55 industrialized countries, responsible for 55% of the greenhouse gas emissions. Russia's signature, however, was driven by possible political gains in other international affairs and economic advantages due to the Kyoto provisions (Henry and Sundstrom 2007). In 2011, however, just like Canada and Japan, Russia announced that it would not accept new commitments. The more recent Paris Agreement was nonetheless signed on 22 April 2016, although it has not been ratified yet.[8]

5.3 Country Characteristics and Environmental Behaviors

After describing the development in these six countries, this section now discusses the magnitude and change of private and public behaviors with a view on the specific circumstances in these countries. Figures 5.1 and 5.2 depict, for this purpose, both behaviors for 1993, 2000, and 2010 and thus also show the period in between, which was left out in Table 5.1.

The discussion of the USA and Canada showed that the importance of environmental topics in politics is strongly influenced by the ruling party and the swing from the liberal to conservative parties and vice versa. Governmental positions and activities of environmental groups, however, can be in an antagonistic relationship. In the USA, for example, environmental movements and activities flourished under governments that opposed ecological measures (Brulle 2000). In line with this idea of

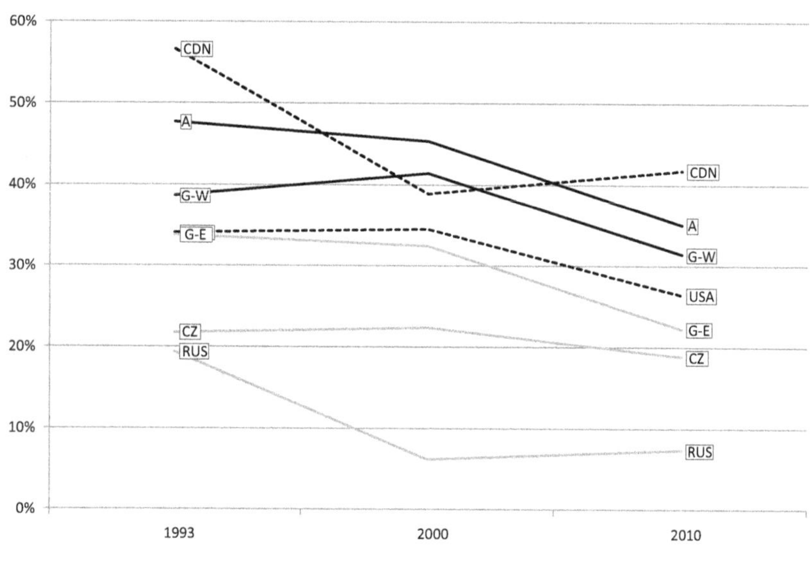

Fig. 5.1 Public behavior over time in six countries. Percentage of respondents who were active in at least one of the four behaviors: signing a petition, taking part in a demonstration, donating money, being member of an environmental group. Data source: ISSP data from 1993, 2000, and 2010.

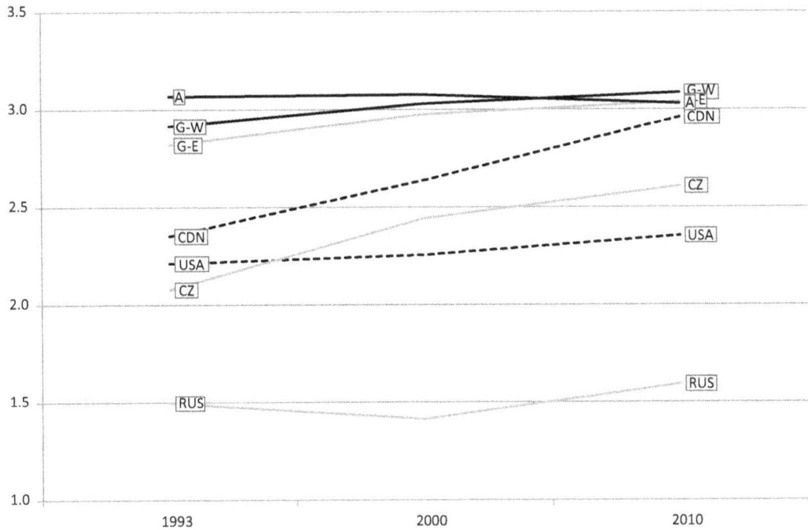

Fig. 5.2 Private behavior over time in six countries. Mean values of an index on private behaviors recycling and cutting back the use of a car (5 indicates 'always' and 1 'never'). Data source: ISSP data from 1993, 2000, and 2010.

opposed forces, public environmental actions hit a low in 2000 in Canada during the liberal party rule and increased again during the conservative rule. In the USA, on the other hand, all ISSP environment surveys were fielded during Democratic presidencies—Clinton I and II and Obama I—and the idea of opposed forces thus cannot be tested as straightforwardly as in Canada.

Comparing the prevalence of private and public behavior in the USA and Canada over the entire period, indicates low levels of environmental actions in the USA, which contradicts the notion of active individual participation (Schofer and Fourcade-Gourinchas 2001). A reason for the low participation in the USA may be the low overall salience of environmental issues compared to other topics, as can be seen in the answers to the relative importance of different societal issues in the 2010 ISSP survey. Respondents were asked to name the most important issue for the country by picking from a list of items such as health care, crime, the environment, and the economy. In the USA, the environment was named only by 4%, while 13% of the Canadians selected it as the main issue.

Considering private behavior Canadians and Americans started around the same level in 1993. Subsequently, Canadians increased their private behaviors and ended up showing similar levels as Germans and Austrians, whereas the Americans did not change much. Here, the national institutionalization of environmentalism might have played a role. In this regard, Hadler and Haller (2011) have shown that governmental regulation of environmental behavior—such as access to recycling—is an important factor in promoting private behaviors.

In line with this argument, Van Nijnatten's (1999) observation of an increasing thrust by the Canadian government to promote private environmental behaviors in the 1990s might explain the increase in this country and the stagnation in the USA, where environmental topics lost importance in the final years of the Clinton presidency and particularly in the subsequent Bush era. Furthermore, the decline in environmental actions in the USA can also be related to the smaller salience of this topic due to the shift of national attention to security issues after the 9/11 terror attacks of 2001 (Hadden and Tarrow 2007).

The low level of activism in the USA results in the surprising finding that the level of public behavior in Germany and Austria is above that of the USA, their classification as countries with a strong emphasis on the state and little on activism notwithstanding. Yet, as pointed out above, the explanation may rather rest within the low salience of the environment as a topic in the USA than in a particularly active populace in the European countries. The high levels of private behavior in Austria and Germany, on the other hand, match their characterizations and may be an effect of the strong institutionalization of this topic and role of the state.

Germany is also interesting, given the different political histories of the western and eastern part. Looking at the two environmental behaviors, differences are minor between these two parts—both are at similar levels and follow similar trends. Yet, the magnitude of private and public behavior of East Germans is lower than in the western part and thus closer to the levels in the other former socialist countries. These differences are certainly influenced by different political opportunities. After all, environmental movements and public support started in Western democracies, which offered more favorable political opportunities than socialist countries (Dalton 1994; Fagan 2004; Gillham 2008; Jones 1993; Manning 1998).

These different opportunities and trajectories are particularly visible when comparing the Czech Republic and Russia. Environmentalism and related movements had a strong period in the first phase of the transition in the early 1990s. Afterwards, however, the Czech Republic became part of the EU and had to implement Western environmental standards. Consequently, we can see that the magnitude of private behavior becomes more similar to the Western countries and surpasses the USA in 2010. In Russia, on the other hand, public behavior drops substantially from 1993 to 2010 and private behavior remains at a low level throughout the entire period.

These country-specific differences must not distract from the overall picture, suggesting that all six countries are characterized by similar trends: a decreasing or overall low level of public behavior and an increasing or overall high level of private behavior. Between these two trends, the decrease of public behavior is surprising given that environmental problems are still omnipresent and that environmental organizations such as Greenpeace are very active. We thus observe an increasing gap between declining public environmental behavior and an increased institutionalization of environmental problems as pointed out by the international organization and social movement literature (Smith 2007; Wapner 2007).

Increasing institutionalization, on the other hand, may have positive effects on private behaviors. Private behavior has accelerated the most in the Czech Republic where the institutionalization of environmental issues has increased substantially since 1993 due to the transition from a former socialist county and the accession to the EU. East Germany, also a former socialist country, became part of a state with an existing tight environmental regime in 1990. Confirming this idea, the level of private behavior in East Germany lies between that of the Czech Republic and West Germany in 1993 and has continued to increase since then.

In sum, this chapter has pointed to a mix of local developments and general trends, which both have an effect on environmental behaviors. The overall trend points to the theme of this book—forces beyond the local level have an influence on environmental behaviors. The following chapter thus will extend the country selection to a total of 16 countries and test for the influence of global ties on public and private behavior, by simultaneously controlling for other country-level factors such as economic development or political history and individual characteristics such as gender and education.

Notes

1. This chapter builds on Hadler and Wohlkönig (2012), Hadler (2013), and Hadler and Kraemer (2016).
2. https://www.nytimes.com/2016/12/07/us/politics/scott-pruitt-epa-trump.html?_r=0 (accessed 27 April 2017).
3. http://www.reuters.com/article/us-climatechange-pact-idUSKCN0XJ0B1 (accessed 17 April 2017).
4. http://www.cbc.ca/news/politics/paris-agreement-trudeau-sign-1.3547822 (accessed 17 April 2017).
5. http://www.salzburg.com/nachrichten/oesterreich/politik/sn/artikel/kyoto-ziel-nicht-erreicht-oesterreich-zahlt-500-mill-euro-90411/ (accessed 17 April 2017).
6. http://orf.at/stories/2348652/(accessed 17 April 2017).
7. https://unfccc.int/paris_agreement/items/9444.php (accessed 17 April 2017).
8. https://unfccc.int/paris_agreement/items/9444.php (accessed 17 April 2017).

References and Further Readings

Andrews, R.N. 2006. *Managing the Environment, Managing Ourselves: A History of American Environmental Policy*. New Haven, CT: Yale University Press.

Brulle, R.J. 2000. *Agency, Democracy, and Nature. The US Environmental Movement from a Critical Theory Perspective*. Cambridge: MIT Press.

Dalton, R.J. 1994. *The Green Rainbow: Environmental Interest Groups in Western Europe*. New Haven, CT: Yale University Press.

EEA. 2017. http://www.eea.europa.eu/soer-2015/countries/czech-republic. Accessed 28 Apr 2017.

Fagan, A. 2004. *Environment and Democracy in the Czech Republic. The Environmental Movement in the Transition Process*. Northampton, MA: Edward Elgar.

Gillham, P. 2008. Participation in the Environmental Movement: Analysis of the European Union. *International Sociology* 23: 67–93.

Hadden, J., and S. Tarrow. 2007. The Global Justice Movement in the United States since Seattle. In *The Global Justice Movement*, ed. Della Porta, 210–231. Boulder, CO: Paradigm Publishers.

Hadler, M. 2013. Environmental Behaviors in a Transatlantic View. *International Journal of Sociology* 43 (4): 87–108.

Hadler, M., and M. Haller. 2011. Global Activism and Nationally Driven Recycling: The Influence of World Society and National Contexts on Public and Private Environmental Behavior. *International Sociology* 26: 315–345.

Hadler, M., and K. Kraemer. 2016. The Perception of Environmental Threats in a Global and European Perspective. In *Green European: Environmental Behaviour and Attitudes in Europe in a Historical and Cross-Cultural Comparative Perspective*, ed. A. Telesiene and M. Gross, 13–30. London and New York: Routledge.

Hadler, M., and P. Wohlkönig. 2012. Environmental Behaviours in the Czech Republic, Austria, and Germany Between 1993 and 2010. Macro-Level Trends and Individual Level Determinants Compared. *Czech Sociological Review* 48: 467–492.

Haller, M., and C. Troy. 2003. Umwelteinstellungen und Umweltverhalten im internationalen Vergleich. *SWS-Rundschau* 43: 81–105.

Harper, Charles. 2012. *Environment and Society. Human Perspectives on Environmental Issues*. 5th ed. Boston, MA: Pearson.

Harrison, K. 2012. A Tale of Two Taxes: The Fate of Environmental Tax Reform in Canada. *Review of Policy Research* 29 (3): 383–407.

Henry, L.A. 2002. Two Paths to a Greener Future: Environmentalism and Civil Society Development in Russia. *Demokratizatsiya* 10 (2): 184–206.

Henry, L.A., and L.M. Sundstrom. 2007. Russia and the Kyoto Protocol: Seeking an Alignment of Interests and Image. *Global Environmental Politics* 7 (4): 47–69.

Hovi, J., D. Sprinz, and G. Bang. 2010. Why the United States Did Not Become a Party to the Kyoto Protocol: German, Norwegian and US Perspectives. *European Journal of International Relations* 7. doi:10.1177/1354066110380964.

Jones, M.E. 1993. Origins of the East German Environmental Movement. *German Studies Review* 16: 235–264.

Manning, N. 1998. Patterns of Environmental Movements in Eastern Europe. *Environmental Politics* 7: 100–134.

Markham, W. 2008. *Environmental Organizations in Modern Germany: Hardy Survivors in the Twentieth Century and Beyond*. New York: Berghahn Press.

Olive, A. 2015. *The Canadian Environment in Political Context*. Toronto: University of Toronto Press.

Pesendorfer, Dieter. 2007. *Paradigmenwechsel in der Umweltpolitik. Von den Anfängen der Umwelt- zu einer Nachhaltigkeitspolitik: Modellfall Österreich?* Wiesbaden: VS Verlag für Sozialwissenschaften.

Peters, G.P., C. Minx, C.L. Weber, and O. Edenhofer. 2010. Growth in Emission Transfers via International Trade from 1990 to 2008. *PNAS* 108 (21): 8903–8908.

Rootes, C. 2003. Environmental Movements. In *The Blackwell Companion to Social Movements*, ed. D.A. Snow, S.A. Soule, and H. Kriesi, 608–640. Oxford: Blackwell.

Schofer, E., and M. Fourcade-Gourinchas. 2001. The Structural Contexts of Civic Engagement: Voluntary Association Membership in Comparative Perspective. *American Sociological Review* 66: 806–828.

Schreurs, M., and E. Papadakis. 2007a. *Historical Dictionary of the Green Movement*. Lanham, MD, Toronto, and Plymouth: Scarecrow Press.

———. 2007b. *Historical Dictionary of the Green Movement*. Lanham, MD: Rowman and Littlefield.

Seifert, F. 2002. *Gentechnik – Öffentlichkeit – Demokratie. Der österreichische Gentechnik-Konflikt im internationalen Kontext*. München: Profil Verlag.

Smith, J. 2007. The Transnational Network for Democratic Globalization. In *The Social Movements Reader*, ed. J. Goodwin and J.M. Jasper, 236–248. Oxford: Wiley-Blackwell.

Stykow, P. 2014. Russland. In *Politische Systeme im Vergleich: Formale und informelle Institutionen im politischen Prozess*, ed. H.J. Lauth, S.303–S.333. Oldenbourg: De Gruyter.

UN. 2015. United Nations Statistical Databases. http://unstats.un.org/unsd/databases.htm

UNEP. 2014. United Nations Environmental Programme. Geodata Portal. http://geodata.grid.unep.ch/

———. 2015. The Emissions Gap Report 2015. United Nations Environmental Programme (UNEP) Nairobi.

Van Nijnatten, D.L. 1999. Participation and Environmental Policy in Canada and the United States: Trends over Time. *Policy Studies Journal* 27 (2): 267–287.

Wapner, P.K. 2007. Transnational Environmental Activism. In *The Social Movements Reader*, ed. J. Goodwin and J.M. Jasper, 226–235. Oxford: Wiley-Blackwell.

World Bank. 2014 and 2015. World Bank Open Data. http://data.worldbank.org/. Accessed at Various Times.

Yanitsky, O.N. 2012. From Nature Protection to Politics: The Russian Environmental Movement 1960–2010. *Environmental Politics* 21 (6): 922–940.

6

Individual Environmentalism

The previous chapter provided an overview of trends in environmental attitudes and behaviors in selected countries and tried to connect these developments with specific political, economic, and environmental problems in these countries. This chapter[1] now turns toward the overall picture and compares the magnitude of and change in environmental behaviors in a larger set of countries. It thus leaves behind the specific idiosyncrasies of countries and rather looks at the overall picture, considering factors such as number of ties to world society, societal affluence, environmental pollution, and so on.

The backdrop of this chapter is the dominant World Society narrative that global cultural models of environmentalism have shaped nation-states, organizations, and actors around the world (Hironaka 2014). In this vein, past research has shown that ties to world society have affected outcomes such as the creation of state parks (Frank et al. 2000), the implementation of environmental protection measures (Schofer and Hironaka 2005), the inclusion of environmentalism in textbooks (Bromley et al. 2011), decelerated deforestation (Shandra et al. 2012), and the lower use of chemical fertilizers and pesticides (Shorette 2012). Similar research on individual environmental concerns has suggested that ties to global ideas via international organizations have positive effects on

M. Hadler, *The Influence of Global Ideas on Environmentalism and Human Rights*,
DOI 10.1057/978-1-137-57440-4_6

individual concerns (Givens and Jorgenson 2013; Hadler and Haller 2011; Marquart-Pyatt 2012; but see: Knight and Messer 2012).

This chapter also aims to empirically test the various conceptualizations of the links between global ideas and individual environmental concerns. It follows the recent push in literature on this topic (Hallett 2010; Pierotti 2013; Velitchkova 2015), but advances these existing studies by considering the theoretical aspects discussed in Chap. 2 and analyzing more than one dimension of individual behaviors. It advances our knowledge on the diffusion process of global cultural models by extending the simple 'exposure to international organizations' model. It does this by considering aspects such as heterogeneous world society in terms of different players (Tsutsui 2004), levels (Vasi 2007), and regional effects (Beckfield 2010; Longhofer and Schofer 2010) and by considering micro-level processes such as an increasing individual agency and individualization (Beck and Beck-Gernsheim 2001), and embedding in conflicting institutional logics.

Following these main thoughts, this chapter is structured as follows: The first section offers a brief summary of 'World Society' theory and proposes several hypotheses that reflect ideas of how to better grasp the macro-micro link within this theory. This is followed by a description of alternative views, and a discussion of the data and analysis strategy employed. The results section starts with a descriptive overview of the changes in public and private behaviors across countries and time, which is followed by various multi-level models. The concluding section discusses the relevance of these findings for the hypotheses and their significance for World Society theory.

6.1 World Society, the Environment, and the Individual

World Society theory conceptualizes the diffusion of global ideas such as environmentalism and human rights principles to nation-states, organizations, and individual actors around the world (Meyer et al. 1997; Ramirez 2012; Schofer et al. 2012). The diffusion of global ideas to organizational and individual actors is described as a complex process that involves a number of different players (Hironaka 2014). World Society's main focus, however, is on the role of international governmental and non-governmental organizations (IGOs and INGOs) and international treaties in this process (Boli and Thomas 1999). Typical institutions with an environmental emphasis are treaties, such as the UN Convention on Long-range Transboundary Air Pollution of 1979 and the Stockholm Convention on Persistent Organic Pollutants of 2001, IGOs such as the UNEP, and INGOs such as Greenpeace or 350.org. The presence of such institutions in a given country, then, is considered 'ties' to the world society and thus to global ideas.

The dominant World Society narrative is that these ties—especially the presence of INGOs—have a positive effect on the diffusion of global ideas, following the basic formula that the impact of international organizations is stronger the more ties are present, the longer they exist, and the more frequently contact occurs between them and local actors (Schofer and Hironaka 2005). Accordingly, research focusing on environmental outcomes has shown positive effects on the creation of parks, state memberships of environmental organizations, the establishment of environmental laws and ministries (Frank et al. 2000), environmental protection (Schofer and Hironaka 2005), and the inclusion of environmentalism in school books (Bromley et al. 2011).

International comparative research on the influence of global ideas on individual attitudes and behaviors has applied World Society's view that 'conformity to evolving external institutions would be most rapid among those actors best linked in to the wider environment, through worlds of associational life (e.g., non-governmental memberships) and cultural life (e.g., through extended schooling)' (Meyer 2010, 13) by assuming that conformity is higher in societies in which more international organizations are present (Boyle 2005). This view can be found in Givens and Jorgenson (2013), Hadler and Haller (2011), and Marquart-Pyatt (2012), who identified positive effects of such exposure on environmental concerns and behaviors of individuals. This consideration leads to the basic expectation that the presence of ties to world society shapes environmental behaviors. This concept will be referred to as 'exposure.'

This basic idea of exposure, however, has been extended in various ways in research on organizational outcomes. A first additional aspect rests on the observation that world society is not a homogenous entity but made up of different actors, levels, and regional densities. Two important and sometimes opposed actors are governments and the civil society. Hafner-Burton and Tsutsui (2005) showed, for example, that the signing of international treaties by governments is often ineffective in reducing human rights violations, whereas the presence of INGOs was effective. Similarly, Tsutsui (2004) did not find any effects of governmental ties on social movements, whereas non-governmental ties had a positive effect. Applying this view on environmentalism, one can expect that environmental behaviors are shaped differently by governmental and non-governmental organizations.

The influence of global ideas and the effects of ties via international organizations can be amplified by nation-states and national polities. Environmentalism, for example, has been institutionalized within nation-states in forms such as environmental ministries, Green parties, laws, and so on (Frank et al. 2000; Hironaka 2014), which, in turn, can influence individual actions such as private environmental behaviors (Hadler and Haller 2011). Following Bromley et al. (2011), who refer to these different implementations as the national institutionalization of environmentalism, individual support thus should be greater in societies with a strong national institutionalization of environmentalism.

National polities, in addition, differ more generally in the location (state versus society focused) and organization (collective versus individual oriented) of political sovereignty and the associated demands on participation (Frank et al. 1995; Schofer and Fourcade-Gourinchas 2001). Liberal states such the USA are an ideal representation of polities with a mix of 'society' and 'individual' foci, whereas Austria represents the opposite poles of 'state' and 'collective' emphasis. Empirically, the effects of these polity types are reflected in the levels of association in old and new social movements (Schofer and Fourcade-Gourinchas 2001), in the low level of environmental concern in Eastern Europe (Haller and Hadler 2008), and also in examples such as the different reactions toward GMO crops in France and Austria. An agreement between NGOs and the government settled the dispute in the latter, whereas violent protests occurred in the former (Seifert 2008). Support for environmental issues thus should be greater in societies that emphasize the role of the individual.

Related to differences across societies, one also needs to consider that the network of international organizations itself is geographically dispersed, characterized by a growing fragmentation and varying regional densities and levels (Beckfield 2010; Vasi 2007). Longhofer and Schofer's (2010) study of the proliferation of environmental associations showed in this regard that national groups appeared mostly after the arrival of international organizations in peripheral countries. Their findings suggest a stronger influence on environmentalism of domestic political structures and movements in Western countries, whereas global ties are more important in peripheral countries. Shorette (2012) also finds regional differences between core, semi-peripheral, and peripheral countries concerning the effects of world society integration on the use of pesticides and fertilizer. The effects, however, are the strongest in core countries, a finding that differs from the previous conclusions. Nonetheless, one can hypothesize that the effects of international organizations differ between core countries and more peripheral areas.

Alternative Theories and Controls

Differences between the core, semi-periphery, and periphery are also discussed in the work of Jorgensen and others, who—in relation to Wallerstein's foundational work—combine World Society ideas with the effects of economic globalization and the relational position of a country (Givens and Jorgenson 2013; Shandra et al. 2009; Shorette 2012). This research points to an ecologically unequal exchange between countries due to economic processes, which has independent effects from international organizations. More developed countries are able to shift the environmental burden to less developed countries, which consequently face negative effects such as biodiversity loss (Shandra et al. 2009), air and water pollution (Jorgenson et al. 2011), and other harms. The effects of these economic processes, however, are not limited to environmental harms, but also shape individuals' environmental concerns. A country's economic background thus needs to be included as a control when studying the effects of international ties.

The economic background and available resources are also relevant for other theories, which need to be considered as alternative explanations: Inglehart's post-materialism thesis with a focus on the impact of values on environmentalism (Inglehart 1995) and the related discussion of the effects of subjective values versus the objective levels of environmental pollution (Best and Mayerl 2013; Brechin 1999), the prosperity thesis which posits that environmental concerns are related to individual and societal wealth (Franzen and Meyer 2010); and social movement literature[2] which focuses on the available resources and mobilizing structures (Gillham 2008; Smith and Wiest 2012). The reference to social movement literature also points to a final control—the level of democratization—that is important in terms of political opportunities (Kriesi et al. 1995; McAdam et al. 2001; Shandra et al. 2012). The analysis also includes pollution measures and democratization indicators as controls in addition to societal affluence.

6.2 Individual-Level Processes

The previous section emphasized the developments at the macro level and how different contextual characteristics such as the number of organizations in a country, the types of ties to the global level, and so on can influence the diffusion of global ideas. This section now turns toward the processes at the individual level and suggests several ways in which individual characteristics influence the reception of global ideas. This discussion has three elements: firstly, more passive individual characteristics that are a proxy for exposure to global ideas; secondly, more active individual actions, considering elements of an inhabited institution approach (Hallett 2010); and, thirdly, the idea of embedding in conflicting logics (Velitchkova 2015).

As for exposure, global scripts have been increasingly incorporated in curricula, as can be seen, for example, in the appearance of environmentalism in school textbooks around the world (Bromley et al. 2011). The educational system therefore is an important accelerating factor, which makes the exposure to global ideas more likely among better-educated individuals. Furthermore, exposure should be also higher among younger cohorts as these developments are rather recent. Finally, individuals in urban areas are also likely to be more exposed due to the presence of international organizations in cities. Education, age, and urban residency are thus frequently used as indicators of exposure to global models (see Hadler and Haller 2011; Pierotti 2013; Velitchkova 2015; Zhou 2013) and educated respondents, younger cohorts, and individuals who live in an urban area thus should display stronger levels of environmental behaviors.

World Society emphasizes that dominant cultural models prescribe an agentic role for individual actors, who often act on behalf of general principles such as human rights and environmentalism (Meyer 2010; Schofer et al. 2012). The implementation of global cultural models is not just a passive process of exposure but also of deliberate actions. Hallett's (2010) ethnography of an elementary school, for example, describes how a newly hired director consciously implemented global scripts and even imposed them on other employees. Similarly, Tsutsui and Shin (2008) have shown how local human rights activists appeal to international organizations to pressure their governments. Respondents who subscribe to an individualistic approach thus should also be more active in environmental behaviors.

The evolving emphasis on the individual, however, is also discussed under the umbrella of individualization (Beck and Beck-Gernsheim 2001). Individualization processes can adversely affect collective political actions and participation, as seen in declining political participation (Dalton 2008) and the retreat from different forms of civil associations (Putnam 2000). The increasing emphasis on individual actorhood could demonstrate the adverse effect that these actors are less inclined to support collective actions. This thought leads to the alternative hypothesis that individualization has negative effects on individuals' environmental behaviors.

Finally, we also need to consider that global cultural models can be contradictory and that individuals can be embedded in micro-institutional settings that oppose certain cultural models. Meyer (2010), in fact, considers individuals exposed to so many different demands that decoupling from some models is inevitable. Negative effects due to the embeddedness in micro-institutional logics of family and religion, with views and practices that contradict global ideas, was shown by Velitchkova (2015) in the case of political violence. As for the environment and support for environmental groups, one can hypothesize that individuals who hold views that contradict the global dominant model of environmentalism will also be more likely not to support environmental organizations.

The above combines the ideas of exposure to World Society and active institution work. However, we also need to acknowledge alternative explanations. The main findings on individual environmental concerns, post-materialist values and wealth (Franzen and Meyer 2010; Inglehart 1995), the social-movement-related aspects of constraints due to biographical availability (McAdam 1986; Xiao and McCright 2012), new social class aspects (Kriesi et al. 1995), and various environmental attitudes and values (Dietz et al. 2005; Dunlap and Jones 2002) are considered as individual-level controls.

6.3 Research Methods

The empirical analyses are based on data collected by the International Social Survey Programme (ISSP). The ISSP environment module was fielded in 1993/94, 2000/01, and 2010/11 (ISSP group 2012). Data for all three waves are available for the following countries: Austria, Bulgaria, Canada, the Czech Republic, Germany (East and West), Great Britain, Israel, Japan, New Zealand, Norway, the Philippines, Russia, Slovenia, Spain, and the USA. The total sample size is 57,874 respondents; the average sample size per country and wave is about 1200 respondents and is representative for the adult population of each country.

Dependent Variables

The ISSP questionnaires contain several repeated questions that are related to environmental behaviors. Public behavior comprises the following items: being a member of an environmental organization and having donated money; taken part in a demonstration; and signed a petition within the last five years. The answer possibilities were simple 'yes' or 'no' options. A respondent who was active in at least one of these behaviors was coded 1. Respondents who were not active in any of these behaviors were coded 0. Private behavior consists of the two items: 'forgoing car driving'; and 'recycling of waste.' Answer categories were 'always,' 'often,' 'sometimes,' and 'never,' with additional options 'no car or driver's license' and 'no recycling available.' The two items were reverse coded (always = 4 and never = 1) and each respondent's replies summed up and divided by the number of valid answers. Respondents who do not own a car are categorized based on their recycling behavior and vice versa.

Independent Variables at the Micro Level

As discussed in the background section, exposure to world societal institutions is more likely among educated, younger, and urban cohorts. Therefore, the items, age (in years), education (six-point scale ranging from no formal education to university degree), and the area of residency (rural to urban, three-point scale) are included.

Individualization is captured by a range of variables indicating retreat from social organizations and emphasis on individual agency: not being a member of a union, a political party, or a religious denomination, not attending church (based on a question about the frequency of church attendance), and a question asking respondents if an individual or the state should be responsible for the implementation of environmental measures. These items were initially included separately in the regression analyses. It turned out that all items, except for the denomination, have a significant negative effect on the dependent variables. They thus were summed up into the index, 'individualization,' and only the effects of this index are reported.

Subscription to conflicting micro-level logics is captured by considering various items on environmental attitudes. Similar to Haller and Hadler (2008) and reconfirmed by exploratory factor analyses, the ISSP items on attitudes are combined into the indices 'Willingness to make sacrifices,' 'Fatalism about environmental issues,' and 'Assessment of environmental risks.' Not to be willing to do anything for the environment, thinking that it is not necessary to do anything at all about the environment, and considering the environment not at risk are indicators of holding views that oppose global cultural models.

The following variables are included as controls. Biographical availability aspects are captured by gender, marital status (married/with partner versus all other groups), and work status (full-time, part-time, student, retiree, homemaker, and a reference group comprising unemployed and disabled). Income, unfortunately, had to be excluded due to limited valid cases. It was not asked in several countries and—adding the missing responses from the countries where it was fielded—is available for only about 48% of the respondents. New social movement aspects are grasped by subjective class (six-point scale ranging from lower to upper class,) and party affiliation (five-point scale left-right plus an additional dummy variable for 'not reported'). Inglehart's (1995) post-materialism is captured by a categorical variable (materialistic, mixed, and post-materialistic) based on the items (1) 'maintain order in the nation,' (2) 'give people more say in government decisions,' (3) 'fight rising prices,' and (4) 'protect freedom of speech.'

Missing answers among the independent variables are treated as follows. When questions were not asked in a substantial number of countries (income, e.g., was not asked in six countries in the first wave) the variable is omitted. If a variable was not asked in one or two countries (class, e.g., in New Zealand in 2010), the values are substituted by the mean value across all countries within the given wave. Finally, missing answers within a country are considered as embedded variables in case more than 2% of the respondents had not answered (see Hardy and Reynolds 2004). This embedded variable method results in two regression parameters: The first parameter indicates the effect of the variable of interest (e.g., education) on the dependent variable. The second parameter indicates the difference between repliers (e.g., education provided) and non-repliers (e.g., education missing) with regard to the dependent variable. Overall, 56,669 cases out of a total sample of 57,874 are included.

Independent Variables at the Macro Level

The core variable of the present analysis, ties to world society, is measured in several ways. First, by the commonly used indicator, the logged count of IGOs and INGOs, defined as the 'number of organizations of which a country or a territory is a member, whether directly or through the presence of members in that country' (UIA 2007, 41). Organizations of type A-D are considered and data are taken from various yearbooks of the Union of International Organizations. Alternatively, a more specific linkage via environmental organizations (EINGO)—provided by David Frank—was considered, but dropped due to a strong multicollinearity with the INGO variable. All INGO and IGO measures are included as the logged number of organizations within a country, but were also tested as density measures that are standardized by the population size as used by Givens and Jorgenson (2013). These density measures resulted in similar outcomes and are thus not reported.

The national institutionalization of environmental measures is captured by an index similar to Bromley et al. (2011) and also labeled 'national environmentalism.' It includes the strength of a Green Party (percentage of popular vote in national parliament/congress elections), the existence of an environmental ministry within the period of four years before the survey wave, the number of international environmental treaties a country has signed at the time of the survey (UNEP 2014), and the percentage of protected land- and/or maritime areas (UNEP 2014). Each sub-index is standardized and the mean value across the four items calculated.

The structural socio-political demands on participation are captured following Schofer and Fourcade-Gourinchas (2001). Canada, Great Britain, New Zealand, and the USA are classified as 'liberal' regimes that demand more participation; Austria, Norway, Spain, and West Germany as 'corporatist/welfare' regimes that demand less participation than the liberal regimes; Bulgaria, the Czech Republic, East Germany, Slovenia, and Russia as post-communist regimes that demand the least participation; and Israel, Japan, and the Philippines as the residual categorization of 'other' regimes. Further, countries are also split into core and semi-peripheral countries, as the used data do not include peripheral countries. Core is defined as being an Organisation for Economic Co-operation and Development (OECD) member before the first survey was conducted in 1993 and excludes all former communist countries of Eastern Europe, Israel, and the Philippines.

As for the social movement aspects, POLITY IV data were considered to capture various aspects of the political opportunity structure. However, the country sample of the present study does not result in sufficient variation among democratization indicators due to a strong overlap with communist history. Thus, only the regime types described before are included. Country-level resources are captured by the level of development (Gross Domestic Product (GDP) per capita in US$1000) based on UN (2014) data. The environmental threats in each country are captured by two pollution measures: logged carbon dioxide (CO_2) output per capita and air pollution as particulate matter (PM10) exposure of the population (World Bank 2015).

Estimation Strategies

The data are hierarchically structured into individuals, countries, and survey waves with different respondents in each wave. This design can be analyzed using country invariant characteristics, such as a communist history as level 3, country characteristics that change over time such as the number of INGOs at a given wave as level 2, and respondents' answers as level 1 (see Deeming and Jones 2015). This approach allows distinguishing between the cross-sectional effects and the longitudinal effects of a variable such as the overall level of wealth and its changes over time.

Given the distributions of the two dependent variables, a logistic model is used for the public behavior and a linear model for the private behavior. The effects of the independent variables thus cannot be compared directly in terms of their strength. We, however, can use the info on significance and direction to determine differences in their influence on the two different behaviors. Further, logistic hierarchical regressions with a logit link (see Guo and Zhao 2000) assume that the level 1 random effect is standard logistic distributed conditional on the upper-level random effects. Variance at level 1 is set to one and thus only the remaining variances at levels 2 and 3 are presented in the results section for the logistic model.

Finally, the number of upper-level observations is at the lower end of the required macro-level units (Stegmueller 2013). Several alternative models were thus tested: (1) regressions with a single upper level of 48 country-wave groups, (2) regressions with fixed country effects and dummy variables for the survey wave, and (3) models in which the sample was also divided into smaller subsamples in order to test for influences from outliers. These alternative models resulted in similar outcomes.

6.4 Public and Private Environmental Behaviors Across Countries and Time

Figures 6.1 and 6.2 show the magnitudes and changes in public and private environmental behaviors in 16 countries since 1993. The countries are sorted according to the prevalence of these behaviors in 1993, with public behavior indicating the percentage of respondents who have been active in at least one public behavior (membership in an environmental group, donating money to an environmental organization, taking part in a protest, signing a petition) and private behavior consisting of the mean value of the index on recycling and car use.

As shown in Fig. 6.1, public behavior is most common in liberal states such as New Zealand and Canada, followed by Austria, Great Britain, and Germany. It is least common in the former socialist countries Russia, Slovenia, and Bulgaria. As for changes over time, public behavior has

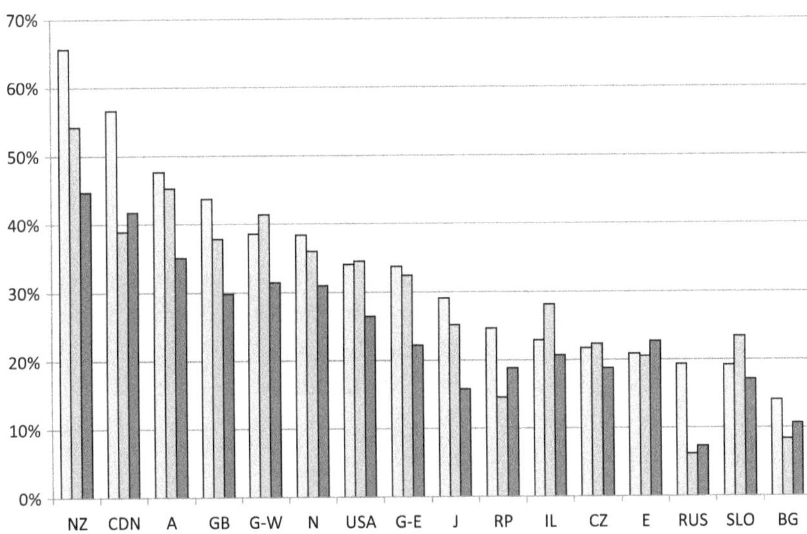

Fig. 6.1 Public environmental behaviors in 1993, 2000, and 2010 across countries. Percentage of respondents who were active in at least one of four different public behaviors. Countries sorted according to the magnitude in 1993. See methods and data section for detailed information. Data source: ISSP (1993), (2000), and (2010)

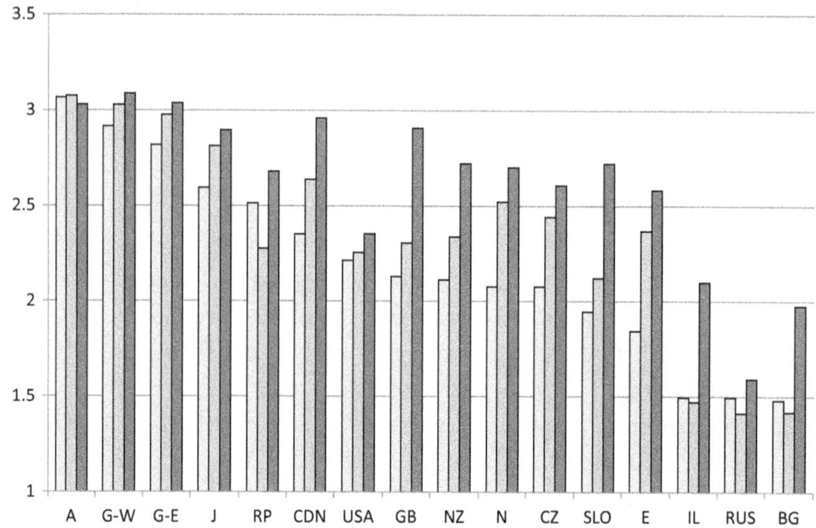

Fig. 6.2 Private environmental behaviors in 1993, 2000, and 2010 across countries. Mean values of an index on private behaviors (5 indicates 'always' and 1 'never'). Countries sorted according to the magnitude in 1993. See methods and data section for detailed information Data source: ISSP (1993), (2000), and (2010)

become less common in countries with an initial high level of public behavior, or remained low in countries with an initially low level. Overall, when comparing 1993 and 2010, a drop in all countries except for Spain can be reported, a small peak in 2000 notwithstanding.

Private behavior, on the other hand, is most common in the corporatist welfare states Austria and Germany and the least common in the two former socialist countries Russia and Bulgaria (see Fig. 6.2). As for changes over time, private behavior has increased the most in countries where it was less common in 1993 and 2000, while it has stagnated at a higher level in more affluent countries.

This initial overview of the two behaviors shows a stark contrast in changes over time, with private behaviors increasing or remaining at a rather high level, whereas public behaviors have declined or remained at a rather low level. These contrarian trends suggest quite diverse effects of global forces, as singular effects of global modals should have resulted in similar trends. Yet, both figures are based only on aggregate figures and cannot grasp the underlying macro-level influences and individual-level dynamics. The following section thus presents the results of various hierarchical models that consider both contextual and individual-level characteristics.

Individual-Level and Country-Level Influences on Environmental Behaviors

After presenting descriptive findings on the magnitude of and changes in public and private environmental behaviors, this section reports the results of various multi-level models. Table 6.1 shows the effects of the micro-level variables and starts with reporting the effects of the indicators on exposure to world society. Education has a significant positive effect on public behaviors, but no effect on private behavior. Age is negatively associated with public behavior and positively linked with private behavior. Urban residency has positive effects on both behaviors. These exposure characteristics are followed by the indicator of individualization, which shows that individualization has a negative effect on both behaviors. As for the subscription to views that oppose global models of environmentalism, the models show that alignment with the globally dominant view of being willing to make sacrifices, considering the environment at risk, and not being fatalistic about possible interventions results in positive effects on both behaviors.

The individual-level control variables indicate some differences between men and women as well as between respondents who are married/with partner versus those without a partner with regard to public behaviors, and various effects of work status characteristics on both behaviors. As for new social movement aspects, negative effects of social class on public behavior and positive effects of the same variable on private behavior can be reported. Furthermore, respondents who have a preference for left parties as well as post-materialists are more likely to show public and private behaviors.

Including various macro-level variables extends these micro-level models. However, in order to cope with the small number of countries and limited statistical power, macro-level variables were initially included as

Table 6.1 The influence of individual-level variables on environmental behaviors

	Public behavior		Private behavior	
	B	Sig.	B	Sig.
World society-related variables:				
Exposure				
Education	0.175	**	0.005	
Age	−0.003	**	0.008	**
Residency (rural-urban)	0.039	**	0.043	**
Individualization (low-high)	−0.147	**	−0.040	**
Environmental views				
Willingness to make sacrifices	0.304	**	0.268	**
Environmental risk perception	0.336	**	0.145	**
Fatalistic attitude	−0.274	**	−0.036	**
Controls:				
Biographical availability				
Married/with partner	−0.026		−0.049	**
Female	−0.025		0.143	**
Work status				
Full-time	0.023		−0.115	**
Part-time	0.131	**	−0.067	**
Unemployed	0.058		0.009	
Student	0.293	**	0.008	
Retired	−0.105	*	−0.014	
Homemaker	−0.115	*	0.010	
Other	Ref		Ref	
New social movement indicators				
Subjective class (low-high)	0.025	*	−0.012	**
Party affiliation (left-right)	−0.098	**	−0.022	**
Post-materialist	0.251	**	0.041	**
Mixed	Ref		Ref	
Materialist	−0.203	**	−0.012	
Remaining variances				
Country level	0.271	**	0.180	**
Wave level	0.079	**	0.076	**
Individual level	NA		0.551	**

Data source: ISSP 1993, 2000, and 2010

Public Behavior: Hierarchical logistic regression. Logit link functions, 16 countries, 48 country*time-points, 56,669 valid cases out of 57,874 total sample size. PQL, RIGLS estimation using the program MlWin. Individual-level variance is set to 1 by default

Private Behavior: Hierarchical linear regression. RIGLs estimation. A total of 52,689 valid cases

Included but not shown in this table: Embedded variables for degree, class, willingness to make sacrifices, and party affiliation not reported. Dummy variable for post-materialism missing is also excluded

Significance: $^{**}p < 0.01$, $^{*}p < 0.05$

single variables in addition to the micro-level model (see Table 6.2). Further, all macro-variables that vary over time, such as the number of INGOs, are included in two ways: as a country-level indicator that captures differences across countries and as a growth indicator that captures effects over time within countries. Variables that do not change over time, but characterize different country groups such as political regimes or core versus semi-periphery, are included as level-three variables.

Considering the effects of various macro-level variables on public behavior, the presence of INGOs—the first variable that captures ties to world society—has a negative effect on public behavior at the wave level and no effects at the overall level. Growing numbers of INGOs are associated with declining public behavior, whereas the overall number of

Table 6.2 The influence of macro-level variables on environmental behaviors when entered in addition to the micro-level variables of Table 6.1

Model	Variables	Level of effect	Public behavior B (SE)	Sig.	Private behavior B (SE)	Sig.
1	INGO (LN)	Country	0.381 (0.497)		0.543(0.232)	*
		Wave	−0.430 (0.128)	**	0.410(0.115)	**
2	IGO (LN)	Country	0.375 (0.494)		0.712(0.387)	
		Wave	−0.261 (0.260)		0.507(0.227)	*
3	National environmentalism	Country	0.516 (0.190)	**	0.463(0.151)	**
		Wave	−0.130 (0.197)		0.267(0.175)	
4	Regime: Liberal	Country	0.978 (0.244)	**	0.255(0.306)	
	Welfare	Country	0.669 (0.245)	**	0.513(0.306)	
	Other	Country	0.142 (0.266)		0.190(0.333)	
	Post-communist	Country	Reference		Reference	
		Wave	(constant)		(constant)	
5	Core	Country	0.701 (0.209)	**	0.413(0.209)	*
		Wave	(constant)			
6	GDP (LN)	Country	0.283 (0.117)	*	0.210(0.100)	*
		Wave	−0.220 (0.098)	*	0.399(0.059)	**
7	CO_2 output per capita (LN)	Country	0.104 (0.197)		−0.041(0.173)	
		Wave	0.808 (0.468)		−0.519(0.438)	
8	PM10 exposure	Country	−0.027 (0.008)	**	−0.016(0.007)	*
		Wave	0.008 (0.004)	*	−0.012(0.003)	**

Data source: ISSP 1993, 2000, and 2010

Each model represents a separate regression that includes the named variable and all variables shown in Table 6.1

See Table 6.1 for the estimation strategies and techniques and additional info on the variables

Significance: ** $p < 0.01$, * $p < 0.05$

INGOs does not matter. IGOs, on the other hand, do not have significant effects on this behavior. National institutionalization matters at the country level, with public behaviors being associated with stronger institutionalization of environmentalism. The regime variables are significant only for public behavior and indicate that public behavior is more common in liberal regimes, followed by welfare regimes. Similarly, public behaviors are more common in core countries. The macro-level control variables show that public behavior is more common in affluent societies, but also decreases with increasing wealth. As for pollution indicators, public behavior is negatively associated with PM10 exposure at the country level and positively at the wave level. CO_2, on the other hand, does not have a significant effect.

Following these models with single macro-level indicators, models were estimated with the goal to combine different macro-variables and to test the effects of INGOs, national environmentalism, regimes, and geopolitical regions when controlled for GDP. The additional regression shows that only GDP remains significant when combined with the two other controls, PM10 and CO_2. As for the substantive variables, these additional models (not shown in the tables) suggest that after controlling for societal wealth, the negative wave effects of INGOs, the country-level effects of national environmentalism, the liberal political regimes, as well as the geopolitical region remain significant.

Public behavior thus declines with increasing numbers of ties to world society, is higher in countries that have a strong institutionalization of environmentalism, in polities which support individual participation, and in core countries. Models that combine more than two macro-level variables exceed the statistical power of the small country sample. Yet, a model combining all factors confirmed, at the very least, the direction of these effects.

Table 6.2 also reports the macro-level findings for private behavior. INGOs have significant effects at both the time and the country level. Private behavior is thus more common in countries with many ties to world society and increases with growing numbers of INGOs. The second variable measuring ties to world society—the number of IGOs—has only one significant effect: growing numbers of IGOs are associated with increasing private behavior. The national institutionalization of environmentalism is significant at the country level, with private behavior being more common in countries with a stronger institutionalization. The regime variables do not have any significant effects on private behavior, whereas differentiating between core and more peripheral countries shows that private behaviors are more prevalent in the core nations.

As for the control variables, private behavior is more common in wealthy societies and increases even further with growing affluence. Among the pollution measures, CO_2 does not have any significant effects. PM10 exposure, on the other hand, is associated with lower levels of private behavior at the country level, whereas increasing levels of PM10 exposure result in less private behavior.

The next step was to estimate advanced models that combine various macro-level variables. The regressions for private behavior show that after controlling for societal wealth, only national environmentalism remains significant, whereas INGOs, regimes, and the geopolitical location lost their significance. Private behavior thus is more common in countries with a strong institutionalization of environmentalism and further increases with growing societal wealth. Given these few significant effects, no additional combinations of variables were tested for private behavior, contrary to public behavior.

In sum, these multi-level models for two environmental behaviors indicate some differences in the determinants of public and private behaviors, but also some similarities. At the micro level, similar effects can be observed for alignment with global cultural models, individualization, and exposure via urban residency. The effects of age and education, however, differ. At the macro level, these models show that the influences on public behavior are more complex than those on private behavior. Private behavior is determined by the national context, whereas global and national forces influence public behavior.

6.5 Discussion

This chapter started from the observation that world society related research into the influence of global cultural models on environmental attitudes and behaviors often subscribes to a simple exposure model that asserts that behaviors and attitudes align with global cultural models when many international organizations are present in a given country. The first descriptive results presented in this paper, however, showed divergent trends in public and private environmental behaviors across countries and over time and thus indicate that these two behaviors are influenced differently by ties to world society.

The effects of these ties, then, were scrutinized in multi-level models that considered both the overall prevalence of international organizations in a given country and the changes in their numbers over time. These analyses found some support for the hypotheses on the positive effects of these ties. Ties to world society have positive effects on private behavior at the national level, but lost their significance once controlled for societal wealth. At the over-time level, negative effects on public behavior can be reported and positive effects on private behaviors. The latter, however, lost their significance once controlled for affluence. The most substantive effect in terms of effects of international organizations is thus the negative association between growing numbers of INGOs and public behaviors—a finding that contradicts the dominant narrative of positive effects.

A possible explanation for this unexpected effect could be the changing nature of organizations, which increasingly focus on support from philanthropists and governments and turn away from individuals (Skocpol 2003). Further, growing numbers of organizations can also have negative effects on protest, as they might have to focus on organizational matters, before they can start garnering public support (Minkoff 1997). Another explanation is the observed shift in protest initiation from formal organizations to loose groups of activists via social media (Wapner 1996), which might also lower the support for traditional formal organizations. These explanations, however, must remain speculation, as this ISSP data do not include any detailed questions on the underlying motivations of our respondents.

More in line with the initial hypotheses, it turned out that distinguishing between governmental and non-governmental international organizations is fruitful, with the latter having more often significant effects than the former. Hafner-Burton and Tsutsui's (2005) findings on the different effects of governmental ratifications of human rights treaties and the presence of INGOs and Tsutsui's (2004) finding on the different effects of INGOs and IGOs on the civil rights movement are thus also reflected in the effects on environmental behaviors. Among these different links, IGOs are the most direct measure of governmental links to world society. World Society scholars, however, also perceive the possibility of indirect effects in the sense that global forces, first, influence national politics, and national politics, in turn, affect actors within the different societies (Longhofer and Schofer 2010). The national institutionalization of environmentalism (Frank et al. 2000; Schofer and Hironaka 2005) has indeed a strong positive effect on both environmental behaviors and thus lends support to this argument.

The analyses also showed that the socio-cultural demands on participation of different polities matter for public behaviors and thus confirmed Schofer and Fourcade-Gourinchas' (2001) view, that activism is particularly high in liberal polities—the group that expects the most individual responsibility from its citizens. The regimes, however, did not have any effects on private behaviors—which were rather influenced by the national institutionalization. This difference suggests that private behaviors are much more driven by national circumstances and regulations than are public behaviors. In addition, the models also show that both behaviors are more common in core countries and thus chime with the findings of Longhofer and Schofer (2010) and Shorette (2012) on differences between geopolitical regions.

Alongside these macro-level processes, the micro-level characteristics: exposure to world society, individualization, and alignment with different cultural environmental models, were considered as well. Among those, only the alignment with cultural models, individualization, and urban residency had uniform effects on both behaviors, whereas education, age, and some control variables had divergent effects.

As for exposure, contrasting the two behaviors has shown that education, which is considered a prime factor in promoting global models (Hadler and Haller 2011; Zhou 2013), has positive effects only on public behavior. In contrast, the prevalence of private behavior is independent from education and rather influenced by availability characteristics. Respondents with availability—older respondents and non-working respondents—are more active in private behavior. This difference suggests that the costliness of a behavior matters for the influence of global models and that individuals may be less inclined to apply global ideas in costly situations—an interpretation that is also supported by Boyle et al.'s (2002) finding that individual attitudes are in line with global models rather than behaviors, as well as by the observation that individuals focus on environmental actions that are socially preferable and that environmental awareness is particularly important in low-cost situations (Diekmann and Preisendoerfer 1998).

The effects of attitudinal alignment with different dimensions of environmentalism support Velitchkova's (2015) view that an embedding in micro-institutional logics is an important factor, in addition to exposure. Respondents who hold views that are in line with the global idea of environmentalism are more active, and—the opposite side of the coin—those who hold conflicting views are less inclined to be active. The regression model presents this finding in a rather smooth fashion, but keeping Hallett's (2010) ethnography of institutional work in mind, we can envision conflicting situations in which individuals are exposed to contrarian pressures and views from their immediate context.

Finally, the effects of individualization warrant further scrutiny. Individualization and the role of individuals were addressed in the background section in two different ways: as a global historical process of prescribing increasing agency to individual actors (Meyer 2010), and as a picture of a general atomization syndrome where detachment from one collective group is accompanied by detachment from other groups (Beck and Beck-Gernsheim 2001). The negative effects of the individualization indicator on both behaviors are rather in line with the second narrative and thus do not bode well with the idea of individuals becoming more agentic (Frank and Meyer 2002; Frank et al. 1995). Both processes, however, seem to work independently from each other, given that an additional test of a cross-level interaction between this individualization trait and the presence of INGOs was not significant.

After discussing the results with regard to the hypotheses posited in the background section, what can be said about the relevance of these findings for world society theory, the existing conceptualization of the macro-micro link, and the way these processes have been studied?

As for diffusion processes, existing research has emphasized influence via exposure (Hadler and Haller 2011; Pierotti 2013; Zhou 2013), embedding in different micro-institutional logics (Velitchkova 2015), and deliberate institutional work of individuals (Hallett 2010). The analysis presented in this chapter supports some of these ideas, but disqualifies others. The effects of the prime exposure factor 'education' were positive for public behavior, but not present for private behavior. This result suggests that the diffusion of global ideas is contingent on the type of behavior. Drawing upon literature of attitudes and costliness, we can suggest that alignment between behavior and the global model 'environmentalism' is stronger in low-cost situations.

This chapter also considered the embeddedness of these micro-level processes in different contexts. The results improve our knowledge in this regard, as they show that global ties are more important for public behavior than for private behavior—a result that chimes with different effects at the micro level, where education influenced public actions, while the private dimension was shaped by availability factors. Analogously, we can suggest that cultural global ideas more often influence public behavior, given that private behavior was shaped by the national implementation of environmentalism and presumably the normative and regulative context of nation-states (Hadler and Haller 2011).

In sum, this chapter has shown that exposure to global ideas is an important factor in the diffusion of global ideas to individuals. Yet, exposure alone is not sufficient. It is paralleled by factors such as attitudinal alignment with global models and works well under certain circumstances, such as low-cost situations and in contexts that are conducive for such behaviors. This chapter, however, was limited to environmental behaviors and a small number of countries. The next chapter thus looks into attitudes toward two different human rights dimensions—xenophobia and homophobia—and scrutinizes the conditions and circumstances that foster or hamper the diffusion of global models to the individual level beyond the area of environmentalism.

Notes

1. This chapter goes back to Hadler and Haller (2011, 2013) and is based on Hadler (2016).

2. The third main approach in social movement literature—framing processes (Benford and Snow 2000)—is also relevant, but has a stronger emphasis on the micro-level and interaction processes, where it is reflected in language such as alignment with global ideas.

References and Further Readings

Beck, U., and E. Beck-Gernsheim. 2001. *Individualization. Institutionalized Individualism and Its Social and Political Consequences*. London: Sage.

Beckfield, J. 2010. The Social Structure of the World Polity. *American Journal of Sociology* 115 (4): 1018–1068.

Benford, R.D., and D.A. Snow. 2000. Framing Processes and Social Movements: An Overview and Assessment. *Annual Review of Sociology* 26 (1): 611–639.

Best, H., and J. Mayerl. 2013. *Postmaterialism and the Environment: A Multilevel Analysis on the 'Objective Problems, Subjective Values' Hypothesis Using the ISSP*. Paper Presented at the WSSF, Montreal, October.

Boli, J., and G.M. Thomas., eds. 1999. *Constructing World Culture: International Nongovernmental Organizations Since 1875*. Stanford, CA: Stanford University Press.

Boyle, E.H. 2005. *Female Genital Cutting: Cultural Conflict in the Global Community*. Baltimore, MD: Johns Hopkins University Press.

Boyle, E.H., B.J. McMorris, and M. Gomez. 2002. Local Conformity to International Norms: The Case of Female Genital Cutting. *International Sociology* 17 (1): 5–33.

Brechin, S.R. 1999. Objective Problems, Subjective Values, and Global Environmentalism: Evaluating the Postmaterialist Argument and Challenging a New Explanation. *Social Science Quarterly* 80 (4): 793–808.

Bromley, P., J.W. Meyer, and F.O. Ramirez. 2011. The Worldwide Spread of Environmental Discourse in History, Civics, and Social Studies Textbooks. *Comparative Education Review* 55: 517–545.

Dalton, R.J. 2008. *Citizen Politics: Public Opinion and Political Parties in Advanced Industrial Democracies*. Washington, DC: CQ Press.

Deeming, C., and K. Jones. 2015. Investigating the Macro Determinants of Self-Rated Health and Well-Being Using the European Social Survey: Methodological Innovations Across Nations and Time. *International Journal of Sociology* 45 (4): 256–285.

Diekmann, A., and P. Preisendoerfer. 1998. Environmental Behavior. Discrepancies Between Aspirations and Reality. *Rationality and Society* 10: 79–102.

Dietz, T., A. Fitzgerald, and R. Shwom. 2005. Environmental Values. *Annual Review of Environmental Resources* 30: 335–372.

Dunlap, R.E., and R.E. Jones. 2002. Environmental Concern: Conceptual and Measurement Issues. In *Handbook of Environmental Sociology*, ed. R.E. Dunlap and W. Michelson, 482–524. Westport, CT: Greenwood Press.

Frank, D.J., A. Hironaka, and E. Schofer. 2000. The Nation-State and the Natural Environment over the Twentieth Century. *American Sociological Review* 65 (1): 96–116.

Frank, D.J., and J.W. Meyer. 2002. The Profusion of Individual Roles and Identities in the Post-War Period. *Sociological Theory* 20: 86–105.

Frank, D., J. Meyer, and D. Miyahara. 1995. Individualist Polity and Professionalization of Psychology. *American Sociological Review* 60: 360–377.

Franzen, A., and R. Meyer. 2010. Environmental Attitudes in Cross-National Perspective: A Multilevel Analysis of the ISSP 1993 and 2000. *European Sociological Review* 26 (2): 219–234.

Gillham, P. 2008. Participation in the Environmental Movement: Analysis of the European Union. *International Sociology* 23: 67–93.

Givens, J., and A.K. Jorgenson. 2013. Individual Environmental Concern in the World Polity: A Multilevel Analysis. *Social Science Research* 42: 418–431.

Guo, G., and H. Zhao. 2000. Multilevel Modeling for Binary Data. *Annual Review of Sociology* 26: 441–462.

Hadler, M. 2016. Individual Action, World Society, and Environmental Change: 1993–2010. *European Journal of Political and Cultural Sociology* 3 (2–3): 341–374.

Hadler, M., and M. Haller. 2011. Global Activism and Nationally Driven Recycling: The Influence of World Society and National Contexts on Public and Private Environmental Behavior. *International Sociology* 26: 315–345.

———. 2013. A Shift from Public to Private Environmental Behavior: Findings from Hadler and Haller (2011) Revisited and Extended. *International Sociology* 28 (4): 484–489.

Hafner-Burton, E., and K. Tsutsui. 2005. Human Rights Practices in a Globalizing World: The Paradox of Empty Promises. *American Journal of Sociology* 110: 1373–1411.

Haller, M., and M. Hadler. 2008. Dispositions to Act in Favor of the Environment: Fatalism and Readiness to Make Sacrifices in a Cross-National Perspective. *Sociological Forum* 23 (2): 281–311.

Hallett, T. 2010. The Myth Incarnate. Recoupling Processes, Turmoil, and Inhabited Institutions in an Urban Elementary School. *American Sociological Review* 75: 52–74.

Hardy, M., and J. Reynolds. 2004. Incorporating Categorical Information into Regression Models. The Utility of Dummy Variables. In *Handbook of Data Analysis*, ed. M. Hardy and A. Bryman, 229–255. Thousand Oaks, CA: Sage.

Hironaka, A. 2014. *Greening the Globe. World Society and Environmental Change.* Cambridge: Cambridge University Press.

Inglehart, R. 1995. Public Support for Environmental Protection: Objective Problems and Subjective Values in 43 Societies. *Political Science and Politics* 15: 57–71.

Jorgenson, A., C. Dick, and J. Shandra. 2011. World Economy, World Society, and Environmental Harms in Less-Developed Countries. *Sociological Inquiries* 81: 53–87.

Knight, K., and B. Messer. 2012. Environmental Concern in Cross-National Perspective: The Effects of Affluence, Environmental Degradation, and World Society. *Social Science Quarterly* 93: 521–537.

Kriesi, H., R. Koopmans, J.W. Duyvendak, and M. Giugni. 1995. *New Social Movements in Western Europe: A Comparative Analysis.* London: UCL Press.

Longhofer, W., and E. Schofer. 2010. National and Global Origins of Environmental Association. *American Sociological Review* 75: 505–533.

Marquart-Pyatt, S.T. 2012. Contextual Influence on Environmental Concern Cross-Nationally: A Multilevel Investigation. *Social Science Research* 41: 1085–1099.

McAdam, D. 1986. Recruitment to High-Risk-Activism. *American Journal of Sociology* 92: 64–90.

McAdam, D., S. Tarrow, and C. Tilly. 2001. *Dynamics of Contention.* Cambridge: Cambridge University Press.

Meyer, J.W. 2010. World Society, Institutional Theories, and the Actor. *Annual Review of Sociology* 36: 1–20.

Meyer, J.W., J. Boli, G.M. Thomas, and F.O. Ramirez. 1997. World Society and the Nation-State. *American Journal of Sociology* 103: 144–181.

Minkoff, D.C. 1997. The Sequencing of Social Movements. *American Sociological Review* 62 (5): 779–799.

Pierotti, R. 2013. Increasing Rejection of Intimate Partner Violence. Evidence of Global Cultural Diffusion. *American Sociological Review* 78 (2): 240–265.

Putnam, R.D. 2000. *Bowling Alone: The Collapse and Revival of American Community*. New York: Simon & Schuster.

Ramirez, F.O. 2012. The World Society Perspective: Concepts, Assumptions, and Strategies. *Comparative Education* 48 (4): 423–439.

Schofer, E., and M. Fourcade-Gourinchas. 2001. The Structural Contexts of Civic Engagement: Voluntary Association Membership in Comparative Perspective. *American Sociological Review* 66: 806–828.

Schofer, E., and A. Hironaka. 2005. The Effects of World Society on Environmental Protection Outcomes. *Social Forces* 84: 25–47.

Schofer, E., A. Hironaka, D. Frank, and W. Longhofer. 2012. Sociological Institutionalism and World Society. In *The New Blackwell Companion to Political Sociology*, ed. E. Amenta, K. Nash, and A. Scott, 57–68. Oxford: Wiley-Blackwell.

Seifert, F. 2008. Einhellige Abschottung und globalisierungskritische Bauernrevolte. Widerstand gegen die grüne Gentechnik in Österreich und Frankreich. *SWS-Rundschau* 48 (4): 485–504.

Shandra, J., L. Esparza, and B. London. 2012. Nongovernmental Organizations, Democracy, and Deforestation: A Cross-National Analysis. *Society and Natural Resources* 25: 251–269.

Shandra, J., C. Leckband, L. McKinney, and B. London. 2009. Ecologically Unequal Exchange, World Polity, and Biodiversity Loss: A Cross-National Analysis of Threatened Mammals. *International Journal of Comparative Sociology* 50: 285–310.

Shorette, K. 2012. Outcomes of Global Environmentalism: Longitudinal and Cross-National Trends in Chemical Fertilizer and Pesticide Use. *Social Forces* 91: 299–325.

Skocpol, T. 2003. *Diminished Democracy: From Membership to Management in American Civic Life*. Norman: University of Oklahoma Press.

Smith, J., and D. Wiest. 2012. *Social Movements in the World System. The Politics of Crisis and Transformation*. New York: Russel Sage Foundation.

Stegmueller, D. 2013. How Many Countries for Multilevel Modeling? A Comparison of Frequentist and Bayesian Approaches. *American Journal of Political Science* 57: 748–761.

Tsutsui, K. 2004. Global Civil Society and Ethnic Social Movements in the Contemporary World. *Sociological Forum* 19 (1): 63–88.

Tsutsui, K., and Hwa-Ji Shin. 2008. Global Norms, Local Activism and Social Movement Outcomes: Global Human Rights and Resident Koreans in Japan. *Social Problems* 55: 391–418.

UIA (Union of International Associations). Various Years. *Yearbook of International Organizations*. Munich, Germany: K.G. Saur.

UNEP. 2014. United Nations Environmental Programme. Geodata Portal. http://geodata.grid.unep.ch/

Vasi, B. 2007. Thinking Globally, Planning Nationally, and Acting Locally. Nested Organisational Fields and the Adoption of Environmental Practices. *Social Forces* 86: 113–137.

Velitchkova, A. 2015. World Culture, Uncoupling, Institutional Logics, and Recoupling: Practices and Self-Identification as Institutional Microfoundations of Political Violence. *Sociological Forum* 30 (3): 698–720.

Wapner, P.K. 1996. *Environmental Activism and World Civic Politics*. Albany: Suny Press.

World Bank. 2014 and 2015. World Bank Open Data. http://data.worldbank.org/. Accessed at Various Times.

Xiao, C., and A. McCright. 2012. A Test of the Biographical Availability Argument for Gender Differences in Environmental Behaviors. *Environment and Behavior*. Published Online Before Print, 9 Aug 2012. doi:10.1177/0013916512453991.

Zhou, M. 2013. Public Support for International Human Rights Institutions: A Cross-National and Multilevel Analysis. *Sociological Forum* 28 (3): 525–548.

7

Individual Attitudes Toward Human Rights

After analyzing the influence of global cultural models on environmental attitudes and behaviors, this chapter now turns toward social tolerance. The underlying idea is to examine whether a different set of attitudes is influenced by global forces in a similar way. In doing so, social tolerance is further split in two distinctive dimensions—tolerance toward respondents of different race and ethnicity—the xenophobia dimension—and tolerance toward respondents of different sexual orientation—the homophobia dimension. This additional division will allow the differentiation of effects of global cultural models on classic and more recent dimensions of human rights.[1]

In the chapter on the international regime, we have discussed in detail, how and why human rights have become important at the international level. The current chapter thus will only summarize the most important historical trends. Yet when it comes to a state's voting behavior in the United Nations, recent polarizations are visible, which were not looked at in the historical regimes and treaties discussion. In particular, the rights of LGBT (lesbian, gay, bisexual, and transgender) groups and sexual orientation have experienced a turn toward more

M. Hadler, *The Influence of Global Ideas on Environmentalism and Human Rights*,
DOI 10.1057/978-1-137-57440-4_7

traditional perceptions in countries such as Russia and various member states of the Organization of Islamic Cooperation (OIC).

The main focus of this book is on the effects of the international level. Yet, international variation in social tolerance has been analyzed using other theories and concepts (see, e.g., Andersen and Fetner 2008; Quillian 1995; Weldon 2006), which need to be considered as alternative explanations. Typically considered are: the effects of affluence and economic security (Inglehart 1990, 1997), group threat and economic struggle (Blumer 1958; Quillian 1995, 1996), exposure to democratic institutions (Rohrschneider 1999; Peffley and Rohrschneider 2003), and cultural differences (Inglehart and Baker 2000). These approaches will be considered as additional factors in this chapter. Yet, as pointed out in the chapters on the theoretical concepts of World Society theory, human rights have played a central role in several World Society related pieces. This chapter also demonstrates that the inclusion of international linkages adds additional insight beyond these alternative views.

The empirical basis of this chapter is three international surveys: the World Values Survey (WVS), European Values Survey, and PEW public opinion data. Social tolerance is measured, first, by the WVS and EVS question on whether a member of various social groups is accepted as a neighbor, with an emphasis on the two distinctive dimensions of prejudice: xenophobia and homophobia. The PEW data, then, are used to capture changing attitudes toward homosexuality and its links to the political decisions of these countries at the international level. The connection between these different sources is that the more advanced analyses will use similar models and independent variables, which will allow us to compare the effects of certain variables across data sources.

In terms of content, this chapter is organized as follows. The next section summarizes the relevant aspects of World Society, including the concept of polarization. Afterwards, alternative views are briefly discussed with an emphasis on theories that consider economic and political factors. The subsequent section describes the data sources, research methods, and variables. The results are presented, first, as an overview of societal differences in intolerance that shows a substantial variation across countries and a decrease of homophobia over time. These descriptive overviews are followed by hierarchical regression models and a concluding discussion, and conclusion section.

7.1 Word Society and Social Tolerance

World Society theory asserts that a network of governmental and non-governmental institutions propels ideas at a global level, and that these ideas influence social organizations and actors within any given nation. A main concern of this international network is human rights and the protection of minorities (Tsutsui 2006). As research has shown that global institutions impact individual actors in such diverse areas as genital mutilation (Boyle 2002) and environmental activism (Hadler and Haller 2011), it seems logical that international organizations also foster social tolerance—especially given that human rights is a key area of emphasis.

A core element of the international regime is human rights (Meyer 2007; Stacy 2009; Tsutsui and Wopitka 2004). Its starting point was the human rights declaration of 1948 that defined certain basic rights such as the right to life; rights in civil and political society; political and religious freedoms; and social, economic, and cultural rights. The initial focus was on characteristics such as race, ethnicity, and religion; more recently, however, the focus has shifted toward characteristics such as sexual preferences, health and illness, criminal histories, and so on. Two non-governmental organizations that reflect these changes are Amnesty International (AI) and Human Rights Watch (HRW). AI was founded in 1961 and focused initially on political prisoners. HRW was founded in 1978 to monitor government compliance with the 1975 Helsinki Accords throughout the Soviet bloc. Both increased their scope, and their current web pages include topics such as health and human rights, HIV/AIDS, LGBT rights and others (see Chap. 4).

The World Society theory considers these governmental and non-governmental networks a carrier of ideas, resources, and power that affects actors around the world (Meyer et al. 1997; Schofer and Hironaka 2005). Interestingly, this global civil society also impacts the behavior of states despite the common governmental practice of a symbolic ratification of treaties (Hafner-Burton and Tsutsui 2005). The influence of world societal ideas, however, is not limited to organizations and governments. Research has also shown that individual actions and behaviors among the general public are affected (Boyle 2002; Hadler and Haller 2011). At a general level, the assumptions are that the more international organizations are present, the longer they exist, and the more contact between them and local actors, the more likely world ideas spill over into the general population (Schofer and Hironaka 2005). More specifically, research has pinpointed several ways of how these ideas work on the ground and are picked up by single individuals. Hallett (2010) describes in an ethnography of an elementary school how a newly hired director implemented global scripts and enforced them on employees. Tsutsui and Shin (2008) discuss how local human rights activists deliberately appeal to international organizations to pressure governments. In a similar vein, Iglic (2010) discusses within a social capital framework that the mere involvement in organizations can increase the exposure to different ideas and consequently promote social tolerance.

Given the world society assertion that local actors acquire rationalized progressive inclinations to the extent that they are linked to international society (Hafner-Burton and Tsutsui 2005; Meyer et al. 1997), and the central role of human rights within this global society, we can expect that the presence of international organizations will foster social tolerance. In a time-comparative view, we can hypothesize that increasing global linkage has a positive effect on social tolerance.

7.2 A Recent Split in World Society

There are, however, additional points to consider. While scholars have indicated that world society is not a homogenous entity but comprises competing ideas and organizations (Hadler and Meyer 2009; Tsutsui 2004), this point has not been analyzed frequently. World Society thus often faces the critique that it fails to recognize heterogeneity and the continuing role of conflict in world society. This shortcoming appears increasingly relevant in the context of laws governing sexuality (Cobb 2014; Eisenstadt 2000). Over the last decade, the question of whether human rights protections should cover sexual orientation and gender identity (SOGI) has become increasingly divisive within international human rights bodies; there have been a number of heated debates and, in 2012, a mass walkout from a UN Human Rights Council Meeting. Frank, Boutcher, and Camp's post-1945 longitudinal study of sodomy laws found that 90% of all changes to sodomy laws between 1945 and 2005 were in a liberalizing direction (involving changes in age of consent laws, lifting or reducing criminal sanctions). Only eight countries increased penalties across the 60-year period and rare illiberal changes, such as Gabon in 1969 and Cameroon in 1972, almost always occurred after regime change and were typically justified as efforts to localize penal codes inherited from colonial powers (Frank et al. 2009, 142). By contrast, legal changes between 2005 and 2014 follow a very different pattern. In the nine years following 2005, more states increased penalties on homosexuality (9) than had increased penalties in the previous 60 years (8), and legal reforms that increased penalties were almost as common as reforms that decreased them.[2]

Trends surrounding regulation of sexuality are much more complex than this simple example suggests and since the main emphasis of this chapter concerns world society's influence on public opinion, the complex relationships between changes in national laws are of less importance. Nevertheless, it is clear that if we would consider those states which recognize gay marriage (19 in 2015, the first of which was the Netherlands in 2001), that prohibit discrimination in employment (62 in 2015, the first national non-discrimination laws were passed by Norway in 1981), or in which homosexual sex is not criminalized (118 in 2015 versus 75 in which it is criminalized), a strong liberalizing trend over the last decade would be visible (ILGA 2016). Conversely, considering countries that have recently criminalized the 'promotion of homosexuality' (Russian Federation, Algeria, Lithuania, and Nigeria), or in which a sub-national jurisdiction increased penalties through introduction of Shari'ah law (e.g. Indonesia and Nigeria), or toughened penalties through uncodified changes in legal practice, further instances of polarization across multiple regions would become visible.

The previous section shows that one cannot assume a single global cultural model or a homogeneous world society. Working within international relations literature that derives from World Society research (Finnemore 1996), Jonathan Symons and Dennis Altman (2015) identify disputation over SOGI rights as a case of 'international norm polarization.' They define this term as the most combative response to attempted changes to international norms: 'a candidate norm is accepted by some states but resisted by others, leading to a period of international disputation between two groups of states in which socializing pressures pull states toward compliance with rival norms' (2015, 61). While the dominant global norm requires states to recognize SOGI as a subject of human rights protection, many states continue to assert an older international norm that sexuality is a legitimate subject of criminal regulation. Many of these anti-SOGI nations also assert that human rights claims can only be valid if they reflect the 'universal traditional values of humankind.' Symons and Altman's account of international polarization seeks to explain the current period of polarization, but does not identify whether it is an interim deviation from a longer-term liberalizing trend, or a secular trend that marks a move toward 'multiple modernities' (Eisenstadt 2000). It seems clear that rival global cultural models have emerged that are associated with each of these international norms. Whereas a 'liberal' global cultural model promotes acceptance of sexual and gender diversity, a 'traditional' global cultural model views homosexuality as undesirable and a legitimate target of criminal law.

Such international divisions over homosexuality have been apparent since debates in 1993 over whether the 'International Lesbian and Gay Association' (ILGA) should be granted official consultative status within the United Nations Economic and Social Council (Sanders 1996). By the early 1990s, a new candidate norm of support for extending human rights protection to SOGI was emerging in parts of Europe and South America. However, as long as the USA sided with conservative nations, disagreements over sexuality remained fairly low profile and the older norm endorsing criminal regulation of homosexuality remained dominant within international society. In 1999, when the UN General Assembly's Third Committee added specific reference to killings on the basis of 'sexual orientation' and (later) 'sexual orientation and gender identity' to a regular resolution condemning extra-judicial killings, the vote was contested, but it did not spark a major international controversy (ISHR 2011, 2). However, when the Obama Administration became an active supporter of SOGI rights in 2009—at a time when the Russian Federation and members of the OIC were organizing in opposition—the disagreement gained a higher international profile. A key turning point occurred in November 2010 when an amendment deleting the reference to 'sexual orientation' from the regular resolution on extra-judicial killings was carried (79 to 70, 17 abstentions and 26 absent: ISHR 2010). In response, the USA organized a rival resolution and pressured a variety of states, including South Africa, Angola, Cape Verde, and Rwanda to vote in support of SOGI rights (ISHR 2011).

If the international community of states is increasingly divided over homosexuality, the same appears to be true of global civil society. On the one hand, major Western NGOs such as Oxfam and AI have increasingly adopted the 'liberal' global cultural model, and NGOs such as ILGA and HRW have become consistent advocates for SOGI and human rights. On the other hand, several global civil society actors—mostly Christian or Islamic religious NGOs—have been active in reinforcing the 'traditional' cultural model (see Bob 2012). For example, since 1997 the World Congress of Families (WCF), a grouping of conservative NGOs, has heavily promoted 'traditional family values' through international conferences and support for anti-homosexual civil society groups. WCF have built strong alliances with political and religious groups in Russia and have promoted anti-homosexual laws internationally (Altman and Symons 2016). In turn, the Russian Federation has supported NGOs and conferences opposing the SOGI human rights agenda in Europe (Horvath 2016).

In sum, this discussion shows that in addition to ties to world society, positions and political decisions of states at the international level may also contribute to public opinion and views on tolerance toward different groups. When analyzing homophobia, using PEW data, the analysis will consider the international position of a country in addition to its ties to world society. Furthermore, and this is the focus of the subsequent section, the analysis will also include several indicators of alternative views and theories.

7.3 Alternative Theories and Controls

A first alternative approach that considers the role of economic development and affluence is Ronald Inglehart's post-materialism thesis (Inglehart 1990, 1997; Inglehart and Baker 2000; Inglehart and Welzel 2005).

This theory asserts that individuals develop post-materialistic values once a sufficient level of economic security is achieved. In particular, experiences throughout socialization have a lasting impact on individuals' values. Those who experience scarcity maintain materialistic values while those whose basic needs were fulfilled are more likely to develop post-materialistic values. Inglehart (1999) considers this economic insecurity conducive to xenophobia and authoritarian attitudes, while postmodernists value diversity, emphasize self-expression, and condemn strict norms regarding sexuality and other absolute rules. Social tolerance, therefore, is considered more common among post-materialists. Inglehart's thesis is also applied to the macro level: post-materialistic values are assumed to be more prevalent in affluent societies. These societies, in turn, show a higher interest in well-being, and a declining emphasis on political, religious, and economic authorities. The role of economic circumstances is also considered within approaches that focus on group threat and competition. Competition theory in the field of ethnic and racial relations points out that ethnic antagonism is more likely when different groups compete for limited resources (Blumer 1958; Olzak 1994; Quillian 1995, 1996). The perceived threat depends on the size of the other group; further, other groups are more likely to be perceived as a threat during periods of economic hardship (Persell et al. 2001).

Alongside the aforementioned political positions at the international level, another set of arguments revolves around the idea of tolerance-fostering effects of democratic systems in general. Experiencing the democratic game of balancing competing interests and acknowledging different views is considered essential in increasing social tolerance, while high levels of autocracy and other repressive governmental forms do not foster tolerant views (Weil 1985). In particular, active participation in the political process (Guérin et al. 2004) and exposure to different ideas are considered tolerance-promoting factors (Peffley and Rohrschneider 2003; Rohrschneider 1999). These assumptions rest on the idea of learned tolerance, with the duration of democracy, and the time of exposure being critical. In situations of abrupt change, however, individual attitudes do not adjust immediately; a lag between the external conditions and individual views can be expected (Homans 1974).

The countries considered in the present chapter have quite different histories (Derbyshire and Derbyshire 2000) and include old democracies such as the USA, Great Britain, and France; countries such as Austria that became stable democracies after World War II; Spain and Portugal, where the authoritarian regimes broke down in the 1970s; and finally, former communist countries of Central and Eastern Europe. Among these changes, the effect of the communist system on tolerance has attracted particular research interest. Overall, the level of social tolerance is lower in the Eastern European countries (Todosijevic and Enyedi 2008). In addition, education has a smaller tolerance-increasing effect in Eastern Europe than in Western Europe (Marquart-Pyatt and Paxton 2007).

These theories have been applied to social tolerance toward individuals of different race and ethnicity (Quillian 1995, 1996) and toward homosexuals (Andersen and Fetner 2008). Therefore, we can expect that affluence will foster tolerance in both dimensions while economic hardship will lower tolerance. The literature on the effects of a democratic political system points toward positive effects on social tolerance. Hence we hypothesize that social tolerance is higher in seasoned democracies and lower in former communist countries. In addition to these factors, the dominant societal religion may be significant as well, because evidence suggests involvement in some religious communities is linked to low acceptance, and also because opinions of religious majorities have a wider cultural influence on non-religious members of national communities (Doebler 2015; Kenneavy 2012; Putnam and Campbell 2010).

7.4 Individual-Level Processes

The individual-level determinants have been discussed in detail in the chapter on the individual actor and, thus, are summarized only briefly here, with an emphasis on their overlap with the alternative theories named in the previous section. Exposure to global cultural models is considered central in World Society theory, but differs across social groups—with education, age, and residency frequently used as proxies for the degree of exposure (see Hadler 2016; Pierotti 2013; Velitchkova 2015; Zhou 2013). Education is important as global cultural models have been increasingly incorporated into curricula, as can be seen, for example, in the appearance of environmentalism in school textbooks around the world (Bromley et al. 2011). Exposure is expected to be higher among younger cohorts as the emergence of global cultural models linking SOGI to human rights is fairly recent, and also among urban dwellers due to the more frequent presence of international organizations in urban areas.

Similar, the post-materialism thesis (Inglehart 1990, 1997) refers to the presence of post-materialistic values, which are more common among urbanites, younger cohorts, as well as educated and economically better-off individuals. Education is also considered within the self-interest approach that emphasizes that individuals develop prejudices toward groups and individuals they compete with (Quillian 1995, 1996). It points out that less qualified individuals, in terms of education or occupation, compete more strongly with other social groups.

The understanding of actors as mere script followers due to exposure has been criticized as overly simplistic since institutions are interpreted, modified, and challenged in local interactions (Hallett 2010; Hallett and Ventresca 2006). For example, Hallett (2010) shows in an ethnography of a public school, how a newly hired principal faces various obstacles and conflicts with other school members when seeking to implement the cultural model of 'accountability.' Similarly, Velitchkova (2015) emphasizes that actors can be part of different micro-level logics, of which some can be opposed to global cultural models, while others align with them. Here, the embedding in religious groups that oppose the global liberal model can be an important decelerator of diffusion processes.

As for other individual-level variables, political attitudes are also identified as relevant factors in influencing social tolerance (Golebiowska 1999; Guérin et al. 2004; Verba et al. 1997). For example, adherents of right-wing parties are more xenophobic. Social psychological theories, in addition, refer to subconscious cognitive processes and emotions (Quillian 1995; Sotelo 2002). The classic approach of Adorno et al. (1950) asserts that the views of adults are a result of their personality orientations developed during childhood. Characteristics of an authoritarian personality are, for example, submissive, uncritical attitudes toward moral authorities, and a tendency to condemn and reject people who violate conventional values. Finally, as for gender, it has been shown that women are more tolerant in terms of attitudes toward minorities and homosexuality (Andersen and Fetner 2008; Quillian 1996; Sotelo 2002; Yang 1997).

7.5 Research Methods

The empirical analysis in this chapter is based on two sets of data, collected within the European/ WVS program and the PEW globalization data. As for the WVS and European Values Study (EVS) data, data from WVS 2 (1989–1993), WVS 4 (1999–2004), WVS 5 (2005–2009), WVS 6 (2010–2014), EVS 2 (1990–1993), EVS 3 (1999–2001), and EVS 4 (2008–2010) are combined with the goal of analyzing data from as many countries as possible for at least three waves. WVS 6 data, however, are only supplemental material in addition to the previous waves. It is available for fewer countries and only used to allow some insight into the trends over time. The multi-level analysis of WVS data thus includes individual-level data from 34 countries (with Great Britain and Northern Ireland considered separately and Iceland and Malta excluded due to lack of macro-level data).

The second data source is the *Global Attitudes Survey* from the Pew Research Center, which is available at www.pewglobal.org. Our analysis includes all countries that fielded this survey in 2013 and in at least one of the 2007 and 2002 waves as national representative samples. These three waves result in a total sample size of 72,374 respondents, with an average sample size of 894 respondents per country and wave. The PEW data are included as it offers a different geographical scope, and due to the fact that is was collected within narrow time windows around crucial events at the international level as far as homophobia is concerned.

Dependent Variables

The dependent variable in the WVS/EVS data is intolerance toward different social groups. Several groups are named in the WVS and EVS surveys and respondents were asked for each group whether or not they would mind having them as a neighbor. The time-comparative nature of this analysis limits it to items asked in all surveys. The following groups are therefore considered: individuals of different race, immigrants/foreign workers, individuals who have AIDS, and homosexuals. Factor analyses reveal two underlying dimensions: (a) 'xenophobia' consisting of the immigrant and race/ethnicity items, and (b) a stereotypical perception of equating AIDS with homosexuality as another dimension. This second dimension will be referred to as 'homophobia' despite AIDS and HIV not being limited to homosexuals. Respondents who dislike one of the groups within a given dimension were classified as intolerant in the attendant dimension.

The PEW data are used to measure intolerance toward *homosexuality*. It is captured by responses to the following statement: 'Which one of these statements comes closest to your opinion, number 1 or number 2? #1—Homosexuality should be accepted by society or #2—Homosexuality should not be accepted by society.' Respondents who selected the 'accepted' option are coded as '0' and all other respondents as '1.' A total of 65.7% of the respondents is classified as intolerant.

Independent Variables at the Micro Level

The following independent individual-level variables are included in the WVS/EVS analyses: Age in years; gender as a dummy variable; education as age when schooling was completed; and income (household) measured in terms of domestic income: none, lower third, middle, or upper third. The size of the hometown is measured by an eight-point scale, with 1 indicating a small village with 2000 or fewer inhabitants and 8 a city with 500,000 or more inhabitants. Post-materialism is based on four common Inglehart items: 'Maintaining order in the nation,' 'Giving people more say in important government decisions,' 'Fighting rising prices,' and 'Protecting freedom of speech.' They are included as a categorical variable with the following groups: materialists, mixed, post-materialists, and not classified. The social-psychological aspect of submissive, uncritical attitudes to authorities is measured by: the yes-no question of whether obedience should be taught to children; political attitudes by the left-right scale ranging from 1 = left to 10 = right; and a categorical variable that records whether respondents did classify themselves.

The PEW data allow for similar independent variables, only the items on post-materialism and on authoritarian attitudes were not included. The PEW-based analysis considers the socio-demographics: age, gender, education, and household income as well as religious denomination. Age is measured in years, gender as a categorical variable, education as highest level of education achieved, and income as household income. Education and income were asked in country specific units and thus standardized by dividing them by the national mean and multiplying by the national standard deviation. In addition, a dummy variable and interaction with the main term is included for respondents who did not report their household income. Religious denomination is captured by a categorical variable including the following groups: Catholic, (other) Christian, Orthodox, Jewish, Muslim, Buddhist, other religions, and no denomination. Other religions is a residual group that includes a variety of different religions, but none of these groups had a substantial number of members in any country.

An embedded variable—a specific type of interaction term—is used for missing cases among our interval-level data (see Hardy and Reynolds 2004). This method results in two regression parameters: the first parameter indicates the main effect for those who have a valid reply (e.g., the effect of education on tolerance) and the second parameter the difference between repliers and non-repliers with regard to the dependent variable (e.g., the difference in tolerance between those who reported their education and those who didn't).

Independent Variables at the Macro Level

Contextual characteristics are derived from sources such as Polity Project (2015), World Bank (2014, 2015), UN (2015), and UIA (various years). GDP (in 1000 US dollars per head), GINI coefficient, national unemployment rate (in %), and the net migration (in immigrants per 1000 people) are included to measure aspects of affluence and competition. The national political characteristics include communist legacy (categorical variable), the duration of democracy (in years since 1945), the autocracy peak since 1945 (the highest value during this period) and duration of EU membership (in years). The linkage to world society is measured by country participation in type A through D international non-governmental organizations (INGO) and separately international governmental organizations (IGO). The UIA (2008, 41) defines country participation as 'the number of organizations of which a country ... is a member, whether directly or through the presence of members in this country.'

The analyses based on the PEW data consider the level of affluence measured in GDP in current US dollars (World Bank 2015), ties to world society measured by the number of INGOs present in a given country (UIA various years), the level of social globalization taken from the KOF index, which measures the spread of information, ideas, and people on a scale from 0 to 100 (Dreher 2006); domestic laws as legal status of homosexuality (legal yes/no), the level of democratization as the polity democracy score from 0 to 10 (Polity Project 2015). The dominant societal religion was captured by the size of the main group or in the case of Canada, Czech Republic, and Korea as the historically dominant religion, since the majority reported not belonging to any church. The main groups are: Catholic, Christian, Orthodox, Jewish, Muslim, and East Asian religions.

We also developed indicators of states' international behavior. A variable on states' voting records in respect to human rights and sexual orientation, based on six separate votes conducted in the UN General Assembly's Third Committee (22 November 2002, 16 November 2010, and 21 December 2010), the Human Rights Council (14 July 2012 and 26 September 2014) and a vote on recognition of same-sex relationships among UN staff in the UN General Assembly's Fifth Committee (24 March 2015). Here, countries that always vote in support of protecting SOGI rights are coded as 1, 0.5 indicates mostly supportive votes, 0 abstain or non-attendance, −5 mostly against, and −1 always against. Alternatively, international affiliation was captured through an indicator which contrasted membership of the UN LGBT Core Group (generally reflective of a close US alliance) with membership in the OIC (a group that includes most Muslim majority states and has actively campaigned against SOGI rights), status as a former communist state (generally reflective of increased influence from the

Russian Federation) and the residual group of states that do not fall into any of these three categories.

Estimation Strategies

All dependent variables are nominal, and the data are hierarchically structured with level 1 consisting of individuals living within countries, and level 2 of countries at different time points. This setup is a longitudinal cross-sectional design that is hierarchically structured. Data, therefore, are analyzed using a multi-level approach with countries as level three, country-wave observations as level two, and individuals within a given wave as level one (see Deeming and Jones 2015; Fairbrother 2014 for this setup). This design allows distinctions between cross-sectional effects and longitudinal effects of variables such as institutional ties and societal wealth. Technically, a macro-level variable's mean value across all three waves has to be included for each country as a level three variable, and its growth (macro-level variable at a given wave minus its country average across all three waves) at level two. Further, given that our dependent variable is dichotomous, a logistic hierarchical regression with a logit link is used (see Guo and Zhao 2000). This type of regression assumes that the level 1 random effect is standard logistic distributed conditional on the upper-level random effects. Upper-level random effects are considered normally distributed. Variance at level 1 equals 1, only the remaining variances at level 2 and 3 are therefore presented in the results section.

7.6 Social Tolerance and Intolerance Across Countries and over Time

Table 7.1 shows the level of intolerance across countries and over time. The countries are sorted according to geographical/geopolitical considerations. The Americas—separated into North and Latin America; West European countries; Central and Eastern European countries with a communist past, including Russia and Turkey; and the Asian countries China and India. The values represent the proportion of respondents that

Table 7.1 Levels of social intolerance and their change over time across countries

	Xenophobia				Homophobia			
	EVS2/WVS2[a]	EVS3/WVS4[a]	WVS5[a]/EVS4	WVS 6	EVS2/WVS2[a]	EVS3/WVS4[a]	WVS5[a]/EVS4	WVS 6
	90–93/89–93	99–01/99–04	05–07/08–10	10–14	90–93/89–93	99–01/99–04	05–07/08–10	10–14
Argentina	0.04[a]	0.08[a]	0.04[a]	0.04	0.45[a]	0.25[a]	0.18[a]	0.14
Chile	0.16[a]	0.14[a]	0.12[a]	0.10	0.65[a]	0.42[a]	0.38[a]	0.32
Mexico	0.26[a]	0.20[a]	0.14[a]	0.16	0.71[a]	0.49[a]	0.35[a]	0.28
Canada	0.08	0.06[a]	0.05[a]	NA	0.35	0.22[a]	0.20[a]	NA
USA	0.15	0.12[a]	0.15[a]	0.16	0.46	0.28[a]	0.31[a]	0.26
Austria	0.22	0.15	0.27	NA	0.51	0.32	0.34	NA
Belgium	0.25	0.19	0.09	NA	0.34	0.21	0.09	NA
Denmark	0.13	0.12	0.07	NA	0.15	0.11	0.06	NA
Finland	0.26	0.18	0.20[a]	NA	0.35	0.29	0.29[a]	NA
France	0.16	0.15	0.39[a]	NA	0.28	0.18	0.38[a]	NA
Ireland	0.08	0.17	0.16	NA	0.48	0.35	0.30	NA
Italy	0.17	0.20	0.17[a]	NA	0.51	0.39	0.39[a]	NA
Netherlands	0.11	0.07	0.12[a]	0.21	0.19	0.11	0.14[a]	0.14
Portugal	0.21	0.08	0.15	NA	0.61	0.36	0.34	NA
Spain	0.14[a]	0.15[a]	0.11[a]	0.10	0.38[a]	0.26[a]	0.19[a]	0.14
Sweden	0.10	0.03	0.03[a]	0.05	0.26	0.09	0.07[a]	0.09
Great Britain	0.13	0.17	0.17[a]	NA	0.40	0.32	0.21[a]	NA
N. Ireland	0.11	0.22	0.20	NA	0.53	0.45	0.28	NA
Belarus	0.25[a]	0.25	0.33	0.40	0.84[a]	0.77	0.72	0.81
Bulgaria	0.49	0.35	0.28[a]	NA	0.76	0.64	0.63[a]	NA
Czech Republic	0.43[a]	0.23	0.37	NA	0.70[a]	0.29	0.44	NA
Estonia	0.26	0.25	0.39	0.43	0.81	0.58	0.58	0.62
Latvia	0.35	0.12	0.26	NA	0.84	0.52	0.53	NA
Lithuania	0.27	0.26	0.33	NA	0.91	0.75	0.78	NA

(continued)

Table 7.1 (continued)

	Xenophobia				Homophobia			
	EVS2/WVS2[a]	EVS3/WVS4[a]	WVS5[a]/EVS4	WVS 6	EVS2/WVS2[a]	EVS3/WVS4[a]	WVS5[a]/EVS4	WVS 6
	90–93/89–93	99–01/99–04	05–07/08–10	10–14	90–93/89–93	99–01/99–04	05–07/08–10	10–14
Poland	0.19	0.30	0.19[a]	0.10	0.79	0.63	0.61[a]	0.46
Romania	0.39	0.31	0.24[a]	0.31	0.80	0.70	0.64[a]	0.62
Russia	0.17[a]	0.15	0.36[a]	0.37	0.86[a]	0.69	0.76[a]	0.76
Slovakia	0.50[a]	0.30	0.23	NA	0.80[a]	0.59	0.46	NA
Slovenia	0.50	0.20	0.22[a]	0.18	0.52	0.49	0.42[a]	0.41
Turkey	0.43[a]	0.42[a]	0.37[a]	0.45	0.94[a]	0.92[a]	0.92[a]	0.90
China	0.19[a]	0.22[a]	0.25[a]	0.18	0.81[a]	0.83[a]	0.84[a]	0.81
India	0.48[a]	0.48[a]	0.58[a]	0.70	0.96[a]	0.49[a]	0.55[a]	0.70
Grand mean	0.24	0.20	0.22	0.25	0.59	0.44	0.42	0.47
Std. deviation	0.14	0.10	0.12	0.17	0.24	0.22	0.23	0.27

Values reflect the proportion of respondents who object at least one group within each dimension; values times 100 equals percentages

Data source: World Values Survey (WVS) and European Values Study (EVS)

[a]Data are taken from WVS, otherwise from EVS

object to at least one group within each dimension. In Argentina, for example, 0.04—this equals 4%—of the respondents in wave 1 are xenophobic and thus do not like individuals of a different race, or immigrants, or both groups as a neighbor. The grand mean at the bottom of Table 7.1 is the average across country mean values.

The overall level of xenophobia is 24% in wave 1, 20% in wave 2, 22% in wave 3 and, thus, rather stable over time. Wave 4 shows again a small increase, this increase, however, is smaller considering that the average in wave 3 would be 23% when looking only at the countries that also take part in wave 4. The level homophobia is greater than that of xenophobia, but shows a substantial decline over time. Intolerance in this dimension was 59% in wave 1, 44% in wave 2, and 42% in wave 3. Again, as with xenophobia, a small increase seems to occur in wave 4, as xenophobia measures 47%. However, when limiting wave 3 to the same countries as in wave 4, the mean value would be 48% in wave 3 and we thus we would see a small decline from wave 3 to wave 4.

Considering differences between country groups, it is found that the level of tolerance differs substantially between the geopolitical areas. Intolerance is rather high in the former communist countries of Eastern and Central Europe. Residents in former communist countries that had a very successful transition such as Slovenia, however, are more tolerant. Social tolerance is also rather high among West Europeans with the French in wave 3 being an exception. Here, the data were collected after the civil unrest of 2005, which consisted of a series of riots by mostly Muslim North African youths in Paris and other French cities. This upheaval seems to have a strong impact on both dimensions with xenophobia jumping from 15% to 39% and homophobia from 18% to 38%. In the Americas, the residents of the USA and Canada are rather tolerant in both dimensions; among the Latin American countries, we see an initial strong homophobia, but over time, tolerance increases. As for the two Asian countries, xenophobia is rather high in India and at an average level in China. Homophobia is extremely high in both countries in wave 1, but drops over time in India, despite a rise at wave 4. The Chinese, however, remain homophobic and exhibit a level of intolerance that was common among the residents of Central and Eastern European countries shortly after the fall of the Iron Curtain. It is very likely, thus, that we are observing an influence of the dominant political system. The overall levels of intolerance and the changes over time also indicate possible influences of the dominant religions. First, homophobia is the most pronounced in Turkey—the only Islamic country in our data. Second, the drop in homophobia seems to be somewhat stronger in the dominant Catholic countries than in other countries.

A Second Look at Homophobia

The previous table presented several findings regarding the differences in social tolerance based on WVS/EVS data. Table 7.2 now presents the attitudes toward homosexuality in a different set of countries, based on PEW data. PEW data have the advantage that the data was collected within a narrow time frame in each wave and that the timing of the waves is aligned with important events at the international level. As mentioned in the introduction of this chapter, a major shift in states' positions toward homosexuality occurred in 2008, when the USA shifted its position to a pro-stance. PEW data were collected before and after this crucial point in time.

Table 7.2 shows the aggregate level of intolerance toward homosexuality in 30 countries over time. The level of intolerance varies between slightly above 10% in some countries to levels of more than 90% in other countries. The most accepting respondents can be found in Spain and Germany, followed by Canada and the Czech Republic. The highest rates of intolerance, on the other hand, can be reported for Pakistan, Jordan, Ghana, Egypt, and Indonesia, where only about 2–3% of the respondents agree that homosexuality should be accepted.

The ranking of these countries matches past findings on the association between indicators such as development, wealth, and democracy and the level of acceptance among the public. Intolerance, in addition, also corresponds with the votes of major international players in the UN. Acceptance is rather high in the USA (60% pro)—a major pro-player at the international level since 2008 and low in Russia (16% pro), where, after a period of liberalization in the immediate post-cold war period, politicians have become international promoters of a traditional

Table 7.2 Levels of homophobia and their change over time across countries

	2002	2007	2013
Argentina	0.34	0.28	0.26
Chile	NA	0.36	0.32
Mexico	0.46	0.40	0.39
Canada	0.31	0.30	0.20
USA	0.49	0.51	0.40
France	0.23	0.17	0.23
Germany	0.17	0.19	0.13
Great Britain	0.26	0.29	0.24
Italy	0.28	0.35	0.26
Spain	NA	0.18	0.12
Czech Republic	0.17	0.17	0.20
Poland	0.60	0.55	0.58
Russia	0.78	0.80	0.84
Ukraine[a]	0.83	0.81	0.85
Indonesia	NA	0.97	0.97
Japan	0.46	0.51	0.46
Malaysia	NA	0.92	0.91
Pakistan	0.91	NA	0.98
Philippines	0.36	NA	0.27
South Korea	0.75	0.82	0.61
Israel	NA	0.62	0.60
Jordan	0.88	0.94	0.97
Lebanon	0.79	0.82	0.82
Palest. T.	NA	0.91	0.96
Turkey	0.78	0.86	0.91
Egypt	NA	0.99	0.97
Ghana	0.96	0.96	0.97
Kenya	0.99	0.97	0.92
S. Africa	0.67	NA	0.68
Uganda	0.96	0.98	0.96
Average (country means)	0.40	0.60	0.62

Values reflect the proportion of respondents who picked answer #1 to the following statement: "Which one of these comes closest to your opinion, number 1 or number 2?" "#1—Homosexuality should be accepted by society or #2—Homosexuality should not be accepted by society"

Source: Pew Global Attitudes Surveys

[a]Ukraine did not participate in 2013. 2011 data were used instead

cultural model. Finally, acceptance is low also among the members of the OIC, with only 2% of Pakistanis reporting acceptance.

As for the changes over time at the aggregate level, intolerance seems to remain rather stable, with the overall average equaling 40% in the first and the last wave and around 39% in the middle wave (41%, 40%, and

42% when considering only the countries that took part in all three waves). This 'stability,' however, hides an increasing polarization between countries, which becomes visible upon closer examination. A detailed look shows that our countries can be roughly divided into two groups: countries with low acceptance that are predominantly members of the United Nations' African and Asia-Pacific regional groups; and countries with a high level of acceptance, which are mostly members of the Latin American and Caribbean States, or Western European and Others UN regional groupings. Considering their political affiliation at the UN level, this split between countries also aligns with their stance toward LGBT issues and their closeness to either the pro-group around the USA or the opposed group around Russia and OIC member states.

Countries also move in opposite directions over time. Comparing the latest wave with the earliest wave, trends toward reduced tolerance are visible in the former socialist countries of Eastern Europe, which accords with our observation of the Russian move toward less acceptance and its possible influence on these countries. Alongside the former socialist countries, Table 7.2 indicates that many of the dominant Muslim countries (and OIC members) are also characterized by a shift toward less acceptance. France, in addition, shows a negative trend from a high base, which is likely influenced by the very divisive national debate surrounding the new Socialist government's efforts to recognize same-sex marriage in 2012. Strong homo-positive shifts, on the other hand, can be found in most Western and Latin American countries and a few countries that were characterized by rapid economic development such as South Korea.

Trends in Social Tolerance

These two overviews indicate a few general trends such as decreasing homophobia in the first waves of the surveys and similarities between countries of certain geopolitical positions or levels of affluence. Alongside these similarities within groups, the data also indicate differences between groups of countries. Both xenophobia and homophobia seems to be high in former communist countries and member countries of the OIC. Furthermore, these differences seem to increase over time as far as homophobia is concerned, which indicates an existing and ongoing polarization. This interpretation, however, is based only on aggregated values and ignores possible differences due to the influence of individual characteristics. The following section thus presents the results of multi-level regressions, which simultaneously estimate the effects of country level and individual-level characteristics on both the initial magnitude of social tolerance and its change over time.

7.7 Individual-Level and Country-Level Influences on Social Intolerance

After presenting first descriptive findings on the level of xenophobia and homophobia in the previous section, Table 7.3 shows the results of multi-level regressions that consider individual-level variables and an indicator of time as independent factors. Given that tolerance is a nominal variable, a logistic regression is used. The B values, therefore, are logit coefficients and their interpretation is more complex than those of a linear regression. Nonetheless, positive and negative effects can be identified easily. The dependent variable is coded 1 = intolerance and 0 = tolerance. Therefore, a positive B value indicates increasing intolerance with increasing values of the independent variable and vice versa. The negative effect of education, therefore, shows that intolerance and hostility decrease with higher education. Table 7.3 also reports the remaining variances at the time and country level. These values can be used as a baseline for measuring the influence of macro-level variables in the subsequent models, as the amount of reduction is indicative of the strength of their effects.

Model 1 and Model 2 are based on WVS/EVS and show the effects of various variables on xenophobia and homophobia. Model 3, then, is based on PEW data and shows the effects of our variables on intolerance toward homosexuality. The selection of independent variables varies slightly across these models due to differences in data availability. Yet, all three models include the same core socio-demographic variables. These core variables show similar effects on social tolerance in all three models, with younger respondents, female respondents, better-educated individuals, and more affluent respondents being more tolerant. As for the

Table 7.3 The individual-level determinants of social intolerance (micro-level part of hierarchical regression)

	Xenophobia (WVS/EVS)		Homophobia (WVS/EVS)		Homophobia (PEW)	
	B	Sig.	B	Sig.	B	Sig.
Wave 2 (Ref = 1)	−0.223	*	−0.690	***	−0.012	
Wave 3 (Ref = 1)	−0.147		−0.759	**	0.078	*
Age (years)	0.007	***	0.010	***	0.015	**
Female	−0.070	***	−0.215	***	−0.245	**
Education (years)	−0.060	***	−0.053	***	−2.484	**
HH Income	Not incl.		Not incl.		−1.129	**
Income						
Low	0.136	***	0.045	***	Not incl.	
Medium	Ref.	***	Ref.	***	Not incl.	
High	−0.066	***	−0.117	***	Not incl.	
Political affiliation (left-right)	0.064	***	0.045	***	Not incl.	
Size of hometown (small-large)	−0.027	***	−0.031	***	Not incl.	
Obedience important	0.178	***	0.233	***	Not incl.	
Post-materialism						
Materialists	0.135	***	0.151	***	Not incl.	
Mixed	Ref.	***	Ref.	***	Not incl.	
Post-materialists	−0.324	***	−0.315	***	Not incl.	
Religion (other Christians)	Not incl.				Reference	
Catholic	Not incl.				−0.371	**
Orthodox	Not incl.				−0.221	**
Jewish	Not incl.				−0.838	**
Muslim	Not incl.				0.482	**
Buddhist	Not incl.				−0.247	**
Other religion	Not incl.				−0.110	

(*continued*)

Table 7.3 (continued)

	Xenophobia (WVS/EVS)		Homophobia (WVS/EVS)		Homophobia (PEW)	
	B	Sig.	B	Sig.	B	Sig.
No affiliation	Not incl.				−0.574	**
Remaining time-level variance	0.160	0.029	0.109		0.028	
Remaining country-level variance	0.292	0.087	0.782		1.339	

Model 1 and Model 2 based on WVS/EVS data. Micro-level part of a hierarchical three level logistic regression; logit values; estimation procedures and settings: first-order linearization, MQL estimation, IGLS; 129,226 individual respondents within 32 countries at three different time points; *p*-values: * = 0.05, ** = 0.01, and *** = 0.001

Model 3 based on PEW data. Hierarchical logistic regression. Logit link functions, 30 countries, 80 country*time points, 65,520 valid cases out of 72,375 total sample size. PQL, RIGLS estimation using the program MlWin. Significance: $**p < 0.01$, $*p < 0.05$

Results for embedded variables—the interaction terms for missing answers—are not reported in this table

variables, which are included in the WVS/EVS models only, respondents who classify themselves at the left side of the political left-right scale are more tolerant than those who see themselves at the right; post-materialists and individuals who consider obedience unimportant are less xenophobic and homophobic; and urban dwellers are more tolerant than respondents from rural areas.

The model based on PEW data also considers religious affiliation. Here, respondents without a religious affiliation and Jews are the most tolerant. The effect of Judaism, however, wanes when other macro-level variables are considered (not shown here, see Hadler and Symons 2017), also because most of the Jewish respondents live in Israel. This model thus, first, indicates a split between respondents who belong to a religious group and those without an affiliation. Further, differentiating between religious groups suggests that Muslims display the lowest acceptance of homosexuality among all religious groups.

Finally, as for the changes over time, WVS data indicate that intolerance toward homosexuals and AIDS decreases over time (1989–2010). PEW data, which consider a more recent period (2002–2013), shows no decline over time. This 'stability' indicated in the PEW data, however, does not hold upon closer examination. As the macro-level models will show, the reason for this stability is that countries moved in different directions in the most recent period, which leads to the effect of a stable mean value despite growing differences between countries.

After looking into the micro-level effects, Table 7.4 reports the findings at the macro level. The models presented in this table are an outcome of several additional tests and models such as comparisons with basic models, in which each macro indicator is initially entered as a single additional variable, afterwards removed, and replaced by a different macro indicator; tests for multicollinearity; exclusion of non-significant variables, etc. Table 7.4 thus does not display indicators of all aspects mentioned in the background section, as non-significant variables are only mentioned in the text.

These background comparisons have resulted in a final model for xenophobia that shows that intolerance is significantly higher in the former socialist countries, whereas ties to world society, societal affluence, and the dummy variable for Turkey fail to be significant when combined with these variables. Similarly, the concluding model for homophobia also indicates that the level of intolerance is greater in the former communist countries and societal affluence does not have a significant effect. It, however, also shows that intolerance is higher in Turkey, and that tolerance increases with growing ties to world society.

The model based on PEW data shows that the effects of ties to INGOs (p = 0.056) and the voting pattern at the UN level ($p < 0.05$) remain substantial at the country level when controlled for other variables. A country's international affiliation is not significant for absolute levels of acceptance; international affiliation, however, interacts significantly with the time variable and shows that these groups of countries moved in opposing directions over time—with respondents in post-communist countries and OIC members becoming less accepting, while respondents in the other countries became more accepting.

Table 7.4 Macro-level determinants of xenophobia (macro-level part of hierarchical regression)

	B	Sig.	Remaining variance at the country level	Remaining variance at the time level
Xenophobia (WVS/EVS)				
All indicators of Table 7.3			0.281**	0.165**
+Turkey	0.663			
GDP	−0.012		0.064**	0.180**
Communist legacy	1.097	**		
Number of INGOs (*1000)	−0.012			
Homophobia (WVS/EVS)				
All indicators of Table 7.3			0.685**	0.110**
+Turkey	2.77	**		
GDP	−0.016		0.194**	0.121**
Communist legacy	0.981	**		
Number of INGOs (*1000)	−0.230	*		
Homosexuality (PEW)				
All indicators of Table 7.3				
Time	−0.252	**	0.424**	0.060**
INGO—country	−0.668	+		
Int'l Affiliation None	Ref			
Int'l Affiliation Pro LBGT	0.417			
Int'l Affiliation Postcom	0.430			
Int'l Affiliation OIC	0.629			
Time * None	Ref			
Time * Pro LBGT	0.049			
Time * Postcom	0.258	*		
Time * OIC	0.621	**		
UN votes	−0.845	**		
Social globalization	−0.027	*		

Models 1 and 2 are based on WVS/EVS data: Macro-level part of a hierarchical three-level logistic regression; logit values; estimation procedures and settings: first-order linearization, MQL estimation, IGLS; 129,226 individual respondents within 32 countries at three different time points; p-values: * = 0.05, ** = 0.01. See Table A1 for the coding of all macro variables

Model 3 is based on PEW data. Not shown in table: wave effects of INGO, UN votes, and globalization as they are not significant. Hierarchical logistic regression. Logit link functions, 30 countries, 80 country*time points, 65,520 valid cases out of 72,375 total sample size. PQL, IGLS estimation using the program MlWin. Significance: $**p < 0.01$, $*p < 0.05$, $+p < 0.1$

All models displayed in this table are extensions of the micro-level model displayed in Table 7.3. Each model includes the named predictors and all micro-level variables shown in the previous table

Domestic criminalization, on the other hand, failed to be significant and was excluded, which is also the case for the control variables societal affluence, and democracy. Only social globalization remained significant among the control variables and indicates that acceptance is higher in more globally linked societies.

7.8 Discussion

This chapter set out to discuss the influence of global ties on xenophobia and homophobia, in order to contrast their effects with those on environmental topics. The underlying idea was that similar effects on the dependent variables would indicate similar ways of diffusion and influence.

The number of ties via INGOs was used as an indicator for links to world society. The basic idea that such links foster social tolerance found strong support in homophobia, whereas INGOs failed to remain significant in xenophobia when combined with other macro-level variables. The stronger effects of ties to world society on homophobia are in line with the idea that such ties should have strong effects on the more recent dimensions of social tolerance. As far as homophobia is concerned, these results allow the expanding of findings from the areas of environmentalism (Frank et al. 2010; Hadler and Haller 2011) and human rights (Tsutsui 2006) to social tolerance.

A closer look at homophobia, however, also indicates some counter movements. A secular global trend toward increasingly liberal regulation of homosexuality depicted by Frank et al. (2009) seems to be paralleled by an emerging countertrend toward tougher legal penalties in some regions, which was not captured in previous literature. This supports claims of polarization in world society (Symons and Altman 2015). In line with these legal trends, the analysis reveals that polarization among states is mirrored in public attitudes. Public opinion across countries is diverging—a trend that is obscured by an overall stable mean value of attitudes to homosexuality between 2002 and 2013. This finding led us to extend existing research on the diffusion of global models to the individual level. Whereas earlier studies presumed the dominance of liberal global models (Boyle et al. 2002; Hadler 2012; Velitchkova 2015), these findings suggest that illiberal world cultural models are also increasingly influential. Further, it appears that the OIC and post-Soviet worlds offer alternatives to liberal world culture.

In order to shed some light on the causes of divergence, the influence of various diffusion promoters discussed within World Society was tested. While, the findings confirm previous studies which find that the density of national-level ties with world society influences individual attitudes (Hadler 2012; Marquart-Pyatt 2012; Zhou 2013), the multi-level models also confirm that a state's alignment with world cultural divisions, as captured by UN voting patterns, also shapes public opinion. Tests of interaction between international group affiliation and time show that patterns of change correlate with states' international alignment. Between 2002 and 2013 there was an ongoing decline in individual acceptance of homosexuality within OIC members; after 2007 as the international debate over SOGI rights gained prominence, we observe a decline in former communist states and continuing high level of acceptance within members of the UN LGBT Core Group.

As for alternative views, political circumstances, economic affluence, dominant societal religion, and social globalization was considered in different models. The most influential characteristic among these variables is communist legacy, which, however, as shown in the PEW models also overlaps with the state's international group affiliation in the UN and the associated voting patterns in SOGI issues. Similarly, the models suggest an effect of religious background. The persistent high level of homophobia in Turkey in WVS/EVS data suggests that the societal cultural and religious background as discussed by Inglehart and Baker (2000) also matters. Yet, Turkey was the only Islamic society in the WVS/EVS data used in this chapter. PEW data, which included more countries with an Islamic background, indicate the same differences. However, when including other macro-level variables—particularly affluence and level of democratization—respondents in dominantly Muslim countries do not differ significantly in social tolerance anymore. We thus need to be careful not to overemphasize the role of religion.

As for the diffusion to the individual level, the findings presented in this chapter offer some possible explanations. In the theory part, however, more specific processes such as political actors deliberately seeking out international organizations (Tsutsui and Shin 2008), and leaders within organizations imposing goals on employees (Hallett 2010) were mentioned. The present chapter is limited in this regard as we do not know much about the activities of our respondents. The regression, however, showed that individuals who do classify themselves on the political left-right scale are more tolerant than those who do not classify themselves. Assuming that individuals who classify themselves also have a greater political awareness supports the idea of deliberate actions of empowered individuals. Further, World Society theory considers education a main factor in promoting global scripts (Ramirez 2006). Education did have a significant effect in all models and correlates with increased social tolerance in the respondents. To further test this assumption, various interactions between education and the presence of international organizations were tested (not reported in the tables).

As for WVS data on xenophobia and homophobia, these interactions show a reinforcing effect of these factors, education has a stronger impact when more international organizations are present. This reinforcing effect supports the idea of education being an important linkage between an individual and the global society. PEW data on homophobia were also used to explicitly test for mediation effects. These more advanced models (not reported in the tables) did not indicate a mediation effect of INGO via education, rather a suppression effect as the effect of education becomes stronger when INGOs are included. World Society scholars can suspect that both INGOs and education measure the same underlying concept—global cultural models.

Notes

1. This chapter is based on Hadler (2012) and Hadler and Symons (2017).
2. States that have increased penalties in the period include Uganda, South Sudan, Burundi, Liberia, Nigeria, and Malawi (Malawi criminalized lesbian sex for the first time).

References and Further Readings

Adorno, T.W., E. Frankel-Brunswick, D. Levinson, and R. Nevitt Sanford. 1950. *The Authoritarian Personality*. New York: Harper.

Altman, D., and J. Symons. 2016. *Queer Wars: The New Global Polarization over Gay Rights*. Cambridge: Polity.

Andersen, R., and T. Fetner. 2008. Economic Inequality and Intolerance: Attitudes Toward Homosexuality in 35 Democracies. *American Journal of Political Science* 52: 942–958.

Blumer, H.M. 1958. Race Prejudice as a Sense of Group Position. *The Pacific Sociological Review* 1: 3–7.

Bob, C. 2012. *The Global Right Wing and the Clash of World Politics*. New York: Cambridge University Press.

Boyle, E.H. 2002. *Female Genital Cutting: Cultural Conflict in the Global Community*. New York: The Johns Hopkins University Press.

Boyle, E.H., B.J. McMorris, and M. Gomez. 2002. Local Conformity to International Norms: The Case of Female Genital Cutting. *International Sociology* 17 (1): 5–33.

Bromley, P., J.W. Meyer, and F.O. Ramirez. 2011. The Worldwide Spread of Environmental Discourse in History, Civics, and Social Studies Textbooks. *Comparative Education Review* 55: 517–545.

Cobb, N. 2014. Rethinking the 'World Polity' Perspective on Global Sodomy Law Reform. In *Handbook of LGBT Communities, Crime, and Justice*, ed. D. Peterson and V. Panfil, 283–310. New York: Springer.

Deeming, C., and K. Jones. 2015. Investigating the Macro Determinants of Self-Rated Health and Well-Being Using the European Social Survey: Methodological Innovations Across Nations and Time. *International Journal of Sociology* 45 (4): 256–285.

Derbyshire, D., and I.D. Derbyshire. 2000. *Encyclopedia of World Political Systems*. Vol. 1 and 2. Armonk, NY: Sharpe Reference.

Doebler, S. 2015. Relationships Between Religion and Two Forms of Homonegativity in Europe – A Multilevel Analysis of Effects of Believing, Belonging and Religious Practice. *PLoS One* 10 (8): e0133538.

Dreher, A. 2006. Does Globalization Affect Growth? Evidence from a New Index of Globalization. *Applied Economics* 38 (10): 1091–1110.

Eisenstadt, S.N. 2000. Multiple Modernities. *Daedalus* 129 (1): 1–29.

Fairbrother, M. 2014. Two Multilevel Modeling Techniques for Analyzing Comparative Longitudinal Survey Datasets. *Political Science Research and Methods* 2: 119–140.

Finnemore, M. 1996. Constructing Norms of Humanitarian Intervention. In *The Culture of National Security: Norms and Identity in World Politics*, ed. P. Katzenstein, 153–178. New York: Columbia University Press.

Frank, D.J., S.A. Boutcher, and B. Camp. 2009. The Reform of Sodomy Laws from a World Society Perspective. In *Queer Mobilizations: LGBT Activists Confront the Law*, ed. S. Barclay, M. Bernstein, and A.-M. Marshall, 123–141. New York: NYU Press.

Frank, D.J., B.J. Camp, and S.A. Boutcher. 2010. Worldwide Trends in the Criminal Regulation of Sex, 1945 to 2005. *American Sociological Review* 75 (6): 867–893.

Golebiowska, E.A. 1999. Gender Gap on Political Tolerance. *Political Behavior* 21: 43–66.

Guérin, D., F. Petry, and J. Créte. 2004. Tolerance, Protest and Democratic Transition: Survey Evidence from 13 Post-Communist Countries. *European Journal of Political Research* 43: 371–395.

Guo, G., and H. Zhao. 2000. Multilevel Modeling for Binary Data. *Annual Review of Sociology* 26: 441–462.

Hadler, M. 2012. The Influence of World Societal Forces on Social Tolerance. A Time Comparative Study of Individual Prejudices in 32 Countries. *The Sociological Quarterly* 53: 211–237.

———. 2016. Individual Action, World Society, and Environmental Change: 1993–2010. *European Journal of Political and Cultural Sociology* 3 (2–3): 341–374.

Hadler, M., and M. Haller. 2011. Global Activism and Nationally Driven Recycling: The Influence of World Society and National Contexts on Public and Private Environmental Behavior. *International Sociology* 26: 315–345.

Hadler, M., and J. Meyer. 2009. Tangled Paths to a World Culture. Contradictory Trends in Attitudes to Globalization. In *Charting the Globe. The International Social Survey Program, 1985–2000*, ed. M. Haller, R. Jowell, and T.W. Smith, 267–280. New York: Routledge.

Hadler, M., and J. Symons. 2017. *Divergent Trends in Homophobia and Its Reflection in Public Attitudes*. Paper Presented at the Annual Meeting of the American Sociological Association, Montreal, Canada, August.

Hafner-Burton, E., and K. Tsutsui. 2005. Human Rights Practices in a Globalizing World: The Paradox of Empty Promises. *American Journal of Sociology* 110: 1373–1411.

Hallett, T. 2010. The Myth Incarnate. Recoupling Processes, Turmoil, and Inhabited Institutions in an Urban Elementary School. *American Sociological Review* 75: 52–74.

Hallett, T., and M.J. Ventresca. 2006. Inhabited Institutions: Social Interactions and Organizational Forms in Gouldner's Patterns of Industrial Bureaucracy. *Theory and Society* 35 (2): 213–236.

Hardy, M., and J. Reynolds. 2004. Incorporating Categorical Information into Regression Models. The Utility of Dummy Variables. In *Handbook of Data Analysis*, ed. M. Hardy and A. Bryman, 229–255. Thousand Oaks, CA: Sage.

Homans, G.C. 1974. *Social Behavior. Its Elementary Forms*. Revised ed. New York: Harcourt Brace Jovanovich.

Horvath, R. 2016. The Reinvention of 'Traditional Values': Nataliya Narochnitskaya and Russia's Assault on Universal Human Rights. *Europe-Asia Studies* 68 (5): 868–892.

Iglic, H. 2010. Voluntary Associations and Tolerance: An Ambiguous Relationship. *American Behavioral Scientist* 53: 717–736.

Inglehart, R. 1990. *Culture Shift in Advanced Industrial Society*. Princeton, NJ: Princeton University Press.

———. 1997. *Modernization and Postmodernization: Cultural Economics, and Political Change in 43 Societies*. Princeton, NJ: Princeton University Press.

———. 1999. Postmodernization Erodes Respect for Authority, but Increases Support for Democracy. In *Critical Citizens. Global Support for Democratic Governance*, ed. P. Norris. Oxford: Oxford University Press.

Inglehart, R., and W.E. Baker. 2000. Modernization, Cultural Change, and the Persistence of Traditional Values. *American Sociological Review* 65: 19–51.

Inglehart, R., and C. Welzel. 2005. *Modernization, Cultural Change, and Democracy*. New York: Cambridge University Press.

International Lesbian, Gay, Bisexual, Trans and Intersex Association (ILGA). 2016. Aengus Carroll and Lucas Paoli Itaborahy. In *State Sponsored Homophobia 2016: A World Survey of Laws: Criminalisation, Protection and Recognition*. Geneva: ILGA.

International Service for Human Rights. 2010. *GA Third Committee Deletes 'Sexual Orientation' from Resolution on Extrajudicial Executions*, November 22. http://archive.today/4wUv. Accessed 15 Mar 2016.

———. 2011. *GA Restores Sexual Orientation into EJEs Resolution, Adopts Key Texts on Death Penalty, Iran, DPRK*, 23 Mar. http://www.ishr.ch/general-assembly/983-ga-reintroduces-sexual-orientation-reference-into-ejes-resolution-adopts-key-texts-on-death-penalty. Accessed 15 Mar 2016.

Kenneavy, K. 2012. Support for Homosexuals' Civil Liberties: The Influence of Familial Gender Role Attitudes Across Religious Denominations. *Social Forces* 90 (4): 1347–1375.

Marquart-Pyatt, S.T. 2012. Contextual Influence on Environmental Concern Cross-Nationally: A Multilevel Investigation. *Social Science Research* 41: 1085–1099.

Marquart-Pyatt, S., and P. Paxton. 2007. In Principle and Practice: Learning Political Tolerance in Eastern and Western Europe. *Political Behavior* 29: 89–113.

Meyer, J.W. 2007. Reflections on Institutional Theories of Organizations. In *Handbook of Organizational Institutionalism*, ed. R. Greenwood, C. Oliver, R. Suddaby, and K. Shalin-Andersson, 788–809. Thousand Oaks, CA: Sage.

Meyer, J.W., J. Boli, G.M. Thomas, and F.O. Ramirez. 1997. World Society and the Nation-State. *American Journal of Sociology* 103: 144–181.

Olzak, S. 1994. *Dynamics of Ethnic Competition and Conflict*. Stanford, CA: Stanford University Press.

Peffley, M., and R. Rohrschneider. 2003. Democratization and Political Tolerance in Seventeen Countries: A Multi-Level Model of Democratic Learning. *Political Research Quarterly* 56: 243–257.

Persell, C.H., A. Green, and L. Gurevich. 2001. Civil Society, Economic Distress and Social Tolerance. *Sociological Forum* 16: 203–230.

Pierotti, R. 2013. Increasing Rejection of Intimate Partner Violence. Evidence of Global Cultural Diffusion. *American Sociological Review* 78 (2): 240–265.

Polity Project. 2015. Polity IV Project Data. http://www.systemicpeace.org/. Accessed 15 Mar 2016.

Putnam, R.D., and D.E. Campbell. 2010. *American Grace: How Religion Unites and Divides Us*. New York: Simon & Schuster.

Quillian, L. 1995. Prejudice as a Response to Perceived Group Threat: Population Composition and Anti-Immigrant and Racial Prejudice in Europe. *American Sociological Review* 60: 586–611.

———. 1996. Group Threat and Regional Change in Attitudes Toward African-Americans. *American Journal of Sociology* 102: 816–860.

Ramirez, F.O. 2006. From Citizen to Person? Rethinking Education as Incorporation. In *The Impact of Comparative Educational Research on Neoinstitutional Theory*, ed. A. Wiseman and D. Baker, 367–387. Oxford: Elsevier Science.

Rohrschneider, R. 1999. *Learning Democracy*. New York: Oxford University Press.

Sanders, D. 1996. Getting Lesbian and Gay Issues on the International Human Rights Agenda. *Human Rights Quarterly* 18 (1): 67–106.

Schofer, E., and A. Hironaka. 2005. The Effects of World Society on Environmental Protection Outcomes. *Social Forces* 84: 25–47.

Sotelo, M.J. 2002. Prejudice Against Gypsies Among Spanish Adolescents. *Patterns of Prejudice* 36: 28–43.

Stacy, H. 2009. *Human Rights for the 21st Century*. Stanford, CA: Stanford University Press.

Symons, J., and D. Altman. 2015. International Norm Polarization: Sexuality as a Subject of Human Rights Protection. *International Theory* 7 (1): 61–95.

Todosijevic, B., and Z. Enyedi. 2008. Political Tolerance in Eastern and Western Europe: Social and Psychological Roots, ISSP 31st Annual Scientific Meeting 2008, July, Paris.

Tsutsui, K. 2004. Global Civil Society and Ethnic Social Movements in the Contemporary World. *Sociological Forum* 19 (1): 63–88.

———. 2006. Redressing Past Human Rights Violations: Global Dimensions of Contemporary Social Movements. *Social Forces* 85: 331–354.

Tsutsui, K., and C. Min Wotipka. 2004. Global Civil Society and the International Human Rights Movement. *Social Forces* 83: 587–620.

Tsutsui, K., and Hwa-Ji Shin. 2008. Global Norms, Local Activism and Social Movement Outcomes: Global Human Rights and Resident Koreans in Japan. *Social Problems* 55: 391–418.

UN. 2015. United Nations Statistical Databases. http://unstats.un.org/unsd/databases.htm

Union of International Associations. 2017. http://www.uia.org/faq. Accessed 11 Apr 2017.

Velitchkova, A. 2015. World Culture, Uncoupling, Institutional Logics, and Recoupling: Practices and Self-Identification as Institutional Microfoundations of Political Violence. *Sociological Forum* 30 (3): 698–720.

Verba, S., N. Burns, and K.L. Schlozman. 1997. Knowing and Caring About Politics: Gender and Political Engagement. *The Journal of Politics* 59: 1051–1072.

Weil, F.D. 1985. The Variable Effects of Education on Liberal Attitudes: A Comparative-Historical Analysis of Anti-Semitism Using Public Opinion Survey Data. *American Sociological Review* 50: 458–474.

Weldon, S.A. 2006. The Institutional Context of Tolerance for Ethnic Minorities: A Comparative, Multilevel Analysis of Western Europe. *American Journal of Political Science* 50: 331–349.

World Bank. 2014 and 2015. World Bank Open Data. http://data.worldbank.org/. Accessed at Various Times.

Yang, A.S. 1997. The Polls – Trends: Attitudes Toward Homosexuality. *Public Opinion Quarterly* 61 (3): 477–507.

Zhou, M. 2013. Public Support for International Human Rights Institutions: A Cross-National and Multilevel Analysis. *Sociological Forum* 28 (3): 525–548.

8

Conclusion

The aim of this book was to look into the association between global ideas and their effects on individual environmentalism and attitudes toward human rights. These two topics were chosen as they display different aspects of international cultural models. Environmentalism has roots that go back to the nineteenth century, but its rise occurred only in the 1960s and 1970s. Human rights also has a long history, but gained wide recognition only with the Universal Declaration of Human Rights after World War II. Afterwards, the scope of human rights increased constantly and now includes topics such as sexual orientation in addition to its initial focus on security and political and civic rights.

The empirical analyses of this book consider attitudes and behaviors that reflect these changes. Xenophobia is related to the initial focus of human rights, homophobia to a recent human rights dimension, and environmentalism to an issue that historically gained salience between the two human rights aspects of xenophobia and homophobia. The analyses of the determinants of these attitudes and behaviors is used to answer the substantive question of this book—to improve the conceptualization of the link between global ideas and individual attitudes and behaviors within the World Society theory.

M. Hadler, *The Influence of Global Ideas on Environmentalism and Human Rights*,
DOI 10.1057/978-1-137-57440-4_8

World Society scholars have looked into organizational outcomes such as the funding of local environmental groups or the establishing of human rights institutions. The effect of global ideas on individuals is less often considered. As pointed out throughout this book, this minor role of the individual is surprising given that World Society has its roots in individual-level theories. After all, scholars such as John Meyer and Brian Rowan used ideas of Schütz, Berger, Luckmann, and others, and applied them to organizations and organizational change.

The focus on organizational change shifted subsequently from specific organizations to the diffusion of global ideas to all sorts of organizations and institutions. Ideas, in this sense, refer mostly to a Western cultural account, with its emphasis on rationality and individualism. As for the diffusion of ideas, research was able to identify various paths and obstacles such as an increasing similarity of organizations—dubbed isomorphism—but also processes of detachment when organizational structures are detached from each other or from overarching ideas—dubbed decoupling (Bromley and Powell 2012; Meyer 2007).

International comparative research on individual behaviors and attitudes has picked up World Society ideas and aims to grasp the impact of global ideas on individual attitudes and behaviors. This research often resorts to a simple exposure idea—assuming that the presence of international organizations enriches an individual's context with associated ideas and so causes changes in attitudes and behaviors.

The empirical chapters in this book, however, showed that public environmental behavior has declined in many countries despite an environmentalism-heavy context. They also indicate that human-rights-related attitudes such as tolerance toward people of a particular sexual orientation have improved in some countries, but declined in others, despite an increasing focus of human rights on LGBT topics. These mixed trends in attitudes and behaviors suggest that the connection between global ideas and individual attitudes and behaviors is more complex than a straightforward exposure effect.

Before discussing the general implications of these results, the following paragraphs offer a brief summary of the findings presented in this book. Considering the different aspects of attitudes and behaviors—xenophobia, homophobia, environmentalism—indicative of different time periods and thus of different degrees of institutionalizations, this overview will show that the effects of ties to world society are the strongest for the most recent dimensions.

8.1 Findings on the Environment

The empirical chapters on environmentalism contrast the effects of ties to the global level on public and private behaviors. As seen in previous research (Hadler and Haller 2011), a strong variation across countries can be observed in private behaviors, while public behavior is more uniform across countries. This finding suggests that private behavior is more strongly influenced by local and national circumstances than by public actions. This supports the 2011 conclusion with regard to public behavior that 'there is a global concept of what constitutes appropriate environmentally sound conduct—given that protection of the environment is one of the most established goals within the global community' (p. 333).

Testing the explicit effects of ties to world society via international organizations, however, results in mixed findings. The earlier cross-sectional analysis (Hadler and Haller 2011) identified positive effects of INGOs and IGOs. The current analysis (Chap. 6, see also Hadler 2016) uses longitudinal data and comes to slightly different results: INGOs and IGOs do have positive effects in explaining the level of environmentalism, their over-time effect, in contrast, is negative on public behavior and positive on private behavior. Several effects, in addition, lose their significance once controlled for other variables.

The most surprising effect of international organizations is the negative association between growing numbers of INGOs and public behaviors. This finding contradicts the idea that a strong focus of the international community on environmental issues plus a growing number of such organizations will lead to positive effects on individual environmentalism. Further, these findings also underscored the idea that the association between environmental attitudes and the global level is more complex than a simple one-way influence.

These contradictory results contravene the emphasis in existing research on influence via exposure. The mix of positive, negative, and non-significant effects, in addition, points to the relevance of the embedding in different micro-institutional logics and associated conflicting pressures (Velitchkova 2015). Education, for example, which is considered the prime factor in promoting global models, has positive effects only on public behavior, whereas private behavior is influenced by biographical availability characteristics. The consistently positive effects of attitudinal alignment with different dimensions of environmentalism, on the other hand, support the view that the micro-institutional logics matter. Respondents who hold views that are in line with the global idea of environmentalism are more active than those who hold conflicting views.

These findings also suggest that the diffusion of global ideas is contingent on the type of behavior. Drawing upon literature on environmental attitudes and costliness, one finds that the results also indicate that the alignment between behavior and the global model, 'environmentalism,' is stronger in low-cost situations. Another reason for the mixed results for global forces could be the nature and trajectory of environmentalism. The analyses of human-rights-related attitudes showed that global forces are more influential in the most recent dimensions of human rights, but do not have any effects on the more established dimension of xenophobia. The non-significant effects of some global ideas on environmentalism thus might indicate that some aspects of environmentalism now are already solidified in the form of national institutions and thus independent from additional influences from the global level.

8.2 Findings on Human Rights

Attitudes toward human rights were considered by analyzing social tolerance in the dimensions of xenophobia and homophobia. The results show that xenophobia—intolerance toward people of a different race or ethnicity—varies less across countries than homophobia—intolerance toward homosexuality. Xenophobia, in addition, is much lower than homophobia, which indicates that tolerance in this dimension is more established. The global model of tolerance had time to shape individual attitudes and xenophobia and thus might be independent from further global influences. In line with such an interpretation, INGOs as an indicator for links to world society failed to remain significant in xenophobia when combined with other macro-level variables. Their effect on homophobia, on the other hand, remained strong and significant. These findings underscore an interpretation that global ideas have stronger effects on more recent dimensions of social tolerance and human rights.

Homophobia, however, showed a strong variation across countries. This variation indicates that homosexuality is contested and not equally accepted across countries. A closer look at homophobia identified some counter trends: A secular global trend toward increasingly liberal regulation of homosexuality depicted by Frank et al. (2009) is accompanied by a trend toward strict legal penalties in some regions and supports claims of polarization in world society (Symons and Altman 2015).

This split is also reflected in a diverging public opinion over time. As for the causes of this divergence, the multi-level models confirm that a state's alignment with world cultural divisions shapes public opinion in the sense that patterns of change correlate with states' international alignment in the United Nations (UN). Between 2002 and 2013 there was an ongoing decline in individual acceptance of homosexuality within Organisation of Islamic Cooperation (OIC) members and in former communist states, and a continuing high level of acceptance within members of the UN LGBT core group.

As for the diffusion to the individual level, education had significant effects in all models and increased the social tolerance of the respondents. This finding is in line with the World Society assumption of education being a main promoter of global ideas. Yet, additional models on homophobia using PEW data did not indicate a mediation effect of INGOs via education, rather a suppression as the effect of education became stronger when INGOs were included. World Society scholars can suspect that both INGOs and education encapsulate the same underlying concept—global cultural models.

This interpretation, however, must remain a guess, given the nature of the data used. The regression models are able to identify covariations between the presence of international organizations, length of education, social tolerance, and the like, but cannot grasp the substantive content of these institutions such as campaign topics, school textbooks, and so on. There is, however, some indirect evidence, as World Society scholars have shown that global cultural models are increasingly included in schoolbooks, which clearly indicates that students are more exposed to such ideas.

8.3 Common Findings, Problems, and Concluding Remarks

The background chapters discuss various concepts and approaches to the influence of institutions on the individual level. In this regard, it is useful to distinguish between different elements of institutions and applying Scott's (2014) distinction of the coercive, normative, and cultural pillar of institutions when discussing the effects of global ideas on environmentalism and attitudes to human rights.

Coercion can be found in environmental institutions in such forms as mandatory recycling programs or the banning of cars from inner cities. This type of enforcement needs to have an authority in the background, which is usually a state or government. The empirical results presented in this book suggest that private environmental behavior is more susceptible to coercion than public behavior—recycling and the reduction of car use are more frequent in countries with a strong institutionalization of environmentalism.

As for human rights, coercion has a different connotation and direction as it focuses on the rights of individuals and aims to limit the power of the state. The analyses in Chap. 7 identify some effects of governmental positions on individuals' social tolerance. The state's position at the UN level is associated with the level of acceptance and the direction of its change in various countries. The domestic criminalization of homosexuality, on the other hand, lost its significant impact on individual attitudes once controlled for other factors.

Normative aspects are reflected in standardization procedures such as labeling of organic produce, fair trade certifications, and similar initiatives. Such certifications do not coerce people to buy environmentally sound products or products that were produced in humane conditions. The purchase of such products is rather associated with values, attitudes, and socioeconomic means. The connection to institutionalism is given in the sense that consumers can 'choose' between these products, but the choices are already pre-given and institutionalized in the different certifications.

As for cultural models, events such as Earth Day or Love Parades are examples of institutionalizations that do not aim to coerce individuals or attempt to standardize aspects of social life. INGOs are often in similar situations, as they do not have the same coercive power as governments. The empirical results showed that INGOs do have effects, nonetheless. They showed the weakest effects on behaviors and the strongest on attitudes in the most recent dimension of human rights. The effects of INGOs seem to be stronger in shaping initial perceptions and wane once the goals are institutionalized into norms and laws.

'Exposure' is considered the main way individuals are influenced by such institutions. Studies applying this idea often include INGOs and similar indicators at the macro level, and individual characteristics such as education at the individual level of multi-level models. They assume that an institution-saturated context—having many INGOs—shapes individual attitudes and behaviors (see, e.g., Givens and Jorgenson 2013; Hadler and Haller 2011). Some studies even go so far that the international level is not included at all in the models, but assumed to have a positive converging effect (see, e.g., Pierotti 2013).

The analyses presented in this book went beyond this simple exposure model and aimed to test the diffusion of global ideas via education and other individual-level characteristics. For this purpose, interactions between education and INGOs as well as tests of mediation were included (see Chap. 7, and also Hadler and Symons 2017). These models show that the positive effect of education and its role as mediator of global ideas cannot be presumed: education had no effect on private environmental behavior and no mediation effects could be identified for homophobia. Yet, it had overall positive effects on public behavior and social tolerance.

These findings show that the assumption of a general exposure effect does not apply. This conclusion should not come as a surprise, given that studies on organizational outcomes and on the structure of world society have already identified various counter trends and conflictual situations. Velitchkova's (2015) ideas on the relevance of the embedding in micro-level institutions and possible conflicting demands thus are an important additional aspect. Accordingly, the present results identified conflicting pressures such as holding anti-environmentalist attitudes or belonging to specific religious groups that result in lower environmentalism or greater homophobia, respectively.

These findings map well onto the ideas of exposure and conflicting situations. Yet, these variables are only proxies, as they do not capture the substantive contents of IGOs, INGOs, education, and other factors. Furthermore, the questionnaires asked respondents only if they support environmental organizations, but not which group they support or why they do so. Similar limitations apply to the social tolerance items. Ideally, the presented analyses should be amended with in-depth interviews and probing studies at the international level. Such attempts were made in areas such as national identity (Fleiß et al. 2009) and social inequality (Hafner-Fink et al. 2017). Such additional interviews would allow finding out whether the individual motivations reflect global cultural frames and where conflicts occur.

Further, when talking about global ideas and cultural models, one also needs to consider that these models can be heterogeneous. In the case of human rights, Chap. 4 mentions the Cairo Declaration, which aims to connect ideas of human rights with the religious principles of the Shari'ah, and also points to increased pressure against international organizations in countries such as Russia. In line with these positions, homophobia increased in OIC member countries and Russia over the last decade. Contrasting the different human rights dimensions, on the other hand, showed that these cleavages are more important in the dimension of homophobia, which is relatively new, than in the dimension of xenophobia, which is more long-standing. In the long run, a conversion in attitudes is thus possible.

The global idea of environmentalism is also contested, especially when economic interests and development issues are concerned. Environmentalism was initially considered a problem of the rich world, but environmental problems such as the depletion of the ozone layer and climate change made the global dimension of environmental problems salient. Accordingly, public action and awareness is rather similar across countries. Yet, the more important dimension of environmental behavior—as far as the impact on the environment is concerned—varies substantially across countries. Here, the results suggest that a strong national institutionalization is an effective means for shaping behaviors.

A final remark is necessary on the difference between organizations, attitudes, and behaviors. The discussion of the international level shows that the global community is increasingly populated with treaties, organizations, and other institutions. Attitudes and behaviors cannot accumulate in the same way. They can align with global ideas but also diverge, as seen in the empirical chapters. Organizations, on the other hand, are inherently inert and tend to stick around, even once their goals are reached—as we know from the 'March of Dimes,' an organization that continued to work even after polio was eradicated. Attitudes and behaviors, on the other hand, are more volatile and react to changes at the domestic and global level. We will see if the trends in environmentalism and attitudes to human rights depicted in this book will continue in the same direction, or react to the recent shifts in global politics toward more conservative views.

References and Further Readings

Bromley, P., and W. Powell. 2012. From Smoke and Mirrors to Walking the Talk: Decoupling in the Contemporary World. *The Academy of Management Annals* 6: 483–530.

Fleiß, J., F. Höllinger, and H. Kuzmics. 2009. Nationalstolz zwischen Patriotismus und Nationalismus? Empirisch-methodologische Analysen und Reflexionen am Beispiel des International Social Survey Programme 2003 'National Identity'. *Berliner Journal für Soziologie* 19 (3): 409.

Frank, D.J., S.A. Boutcher, and B. Camp. 2009. The Reform of Sodomy Laws from a World Society Perspective. In *Queer Mobilizations: LGBT Activists Confront the Law*, ed. S. Barclay, M. Bernstein, and A.-M. Marshall, 123–141. New York: NYU Press.

Givens, J., and A.K. Jorgenson. 2013. Individual Environmental Concern in the World Polity: A Multilevel Analysis. *Social Science Research* 42: 418–431.

Hadler, M. 2016. Individual Action, World Society, and Environmental Change: 1993–2010. *European Journal of Political and Cultural Sociology* 3 (2–3): 341–374.

Hadler, M., and M. Haller. 2011. Global Activism and Nationally Driven Recycling: The Influence of World Society and National Contexts on Public and Private Environmental Behavior. *International Sociology* 26: 315–345.
Hadler, M., and J. Symons. 2017. *Divergent Trends in Homophobia and Its Reflection in Public Attitudes*. Paper Presented at the Annual Meeting of the American Sociological Association, Montreal, Canada, August.
Hafner-Fink, M., M. Hadler, F. Höllinger, and S. Uhan. 2017. *Probing the ISSP Social Inequality Module. Results of a Study in Austria and Slovenia*. Paper Presented at the Annual Meeting of the ISSP, Lausanne, CH, April.
Meyer, J.W. 2007. Reflections on Institutional Theories of Organizations. In *Handbook of Organizational Institutionalism*, ed. R. Greenwood, C. Oliver, R. Suddaby, and K. Shalin-Andersson, 788–809. Thousand Oaks, CA: Sage.
Pierotti, R. 2013. Increasing Rejection of Intimate Partner Violence. Evidence of Global Cultural Diffusion. *American Sociological Review* 78 (2): 240–265.
Scott, W.R. 2014. *Institutions and Organizations: Ideas. Interests, and Identities*. 4th ed. Thousand Oaks, CA: Sage.
Symons, J., and D. Altman. 2015. International Norm Polarization: Sexuality as a Subject of Human Rights Protection. *International Theory* 7 (1): 61–95.
Velitchkova, A. 2015. World Culture, Uncoupling, Institutional Logics, and Recoupling: Practices and Self-Identification as Institutional Microfoundations of Political Violence. *Sociological Forum* 30 (3): 698–720.

References and Further Readings

Abramowitz, A., and K. Saunders. 2008. Is Polarization a Myth? *The Journal of Politics* 70 (2): 542–555.

Adorno, T.W., E. Frankel-Brunswick, D. Levinson, and R. Nevitt Sanford. 1950. *The Authoritarian Personality*. New York: Harper.

Almond, G.A., and S. Verba. 1965. *The Civic Culture: Political Attitudes and Democracy in Five Nations*. Boston, MA: Little, Brown.

Altman, D., and J. Symons. 2016. *Queer Wars: The New Global Polarization over Gay Rights*. Cambridge: Polity.

Amnesty. 2017. https://www.amnesty.org/. Accessed 19 Apr 2017.

Andersen, R., and T. Fetner. 2008. Economic Inequality and Intolerance: Attitudes Toward Homosexuality in 35 Democracies. *American Journal of Political Science* 52: 942–958.

Andrews, R.N. 2006. *Managing the Environment, Managing Ourselves: A History of American Environmental Policy*. New Haven, CT: Yale University Press.

Appadurai, A. 1996. *Modernity at Large: Cultural Dimensions of Globalization*. Minneapolis: University of Minnesota Press.

Asal, V., U. Sommer, and P.G. Harwood. 2013. Original Sin. A Cross-National Study of the Legality of Homosexual Acts. *Comparative Political Studies* 46 (3): 320–351.

Ayoub, P.M. 2014. With Arms Wide Shut: Threat Perception, Norm Reception, and Mobilized Resistance to LGBT Rights. *Journal of Human Rights* 13 (3): 337–362.

M. Hadler, *The Influence of Global Ideas on Environmentalism and Human Rights*,
DOI 10.1057/978-1-137-57440-4

———. 2015. Contested Norms in New-Adopter States: International Determinants of LGBT Rights Legislation. *European Journal of International Relations* 21 (2): 293–322.
———. 2016. *When States Come Out*. Cambridge: Cambridge University Press.
Bail, C.A. 2008. The Configuration of Symbolic Boundaries Against Immigrants in Europe. *American Sociological Review* 73 (1): 37–59.
Beck, U. 1986. *Risikogesellschaft. Auf dem Weg in eine andere Moderne*. Frankfurt am Main: Suhrkamp.
———. 1993. *Die Erfindung des Politischen*. Frankfurt am Main: Suhrkamp.
Beck, U., and E. Beck-Gernsheim. 2001. *Individualization. Institutionalized Individualism and Its Social and Political Consequences*. London: Sage.
Beckfield, J. 2010. The Social Structure of the World Polity. *American Journal of Sociology* 115 (4): 1018–1068.
Bell, D. 1976. *The Cultural Contradictions of Capitalism*. New York: Basic Books.
Benford, R.D., and D.A. Snow. 2000. Framing Processes and Social Movements: An Overview and Assessment. *Annual Review of Sociology* 26 (1): 611–639.
Berger, P.L. 1998. Die vier Gesichter der globalen Kultur. *Europäische Rundschau* 26: 105–113.
Berger, P.L., and T. Luckmann. 1966. *The Social Construction of Reality: A Treatise in the Sociology of Knowledge*. Garden City, NY: First Anchor.
Best, H., and J. Mayerl. 2013. *Postmaterialism and the Environment: A Multilevel Analysis on the 'Objective Problems, Subjective Values' Hypothesis Using the ISSP*. Paper Presented at the WSSF, Montreal, October.
Blumer, H.M. 1958. Race Prejudice as a Sense of Group Position. *The Pacific Sociological Review* 1: 3–7.
Bob, C. 2012. *The Global Right Wing and the Clash of World Politics*. New York: Cambridge University Press.
Boli, J., and G.M. Thomas. 1997. World Culture in the World Polity: A Century of International Non-governmental Organization. *American Sociological Review* 62: 171–190.
———, eds. 1999. *Constructing World Culture: International Nongovernmental Organizations Since 1875*. Stanford, CA: Stanford University Press.
Boyle, E.H. 2002. *Female Genital Cutting: Cultural Conflict in the Global Community*. New York: The Johns Hopkins University Press.
———. 2005. *Female Genital Cutting: Cultural Conflict in the Global Community*. Baltimore, MD: Johns Hopkins University Press.
Boyle, E., and A. Corl. 2010. Law and Culture in a Global Context: Interventions to Eradicate Female Genital Cutting. *Annual Review of Law and Social Science* 6: 195–215.

Boyle, E.H., B.J. McMorris, and M. Gomez. 2002. Local Conformity to International Norms: The Case of Female Genital Cutting. *International Sociology* 17 (1): 5–33.

Boyle, E.H., and J.W. Meyer. 1998. Modern Law as a Secularized and Global Model: Implications for the Sociology of Law. *Soziale Welt* 49: 213–232.

Boyle, E.H., K. Minzee, and W. Longhofer. 2015. Abortion Liberalization in World Society, 1960–2009. *American Journal of Sociology* 121 (3): 882–913.

Brechin, S.R. 1999. Objective Problems, Subjective Values, and Global Environmentalism: Evaluating the Postmaterialist Argument and Challenging a New Explanation. *Social Science Quarterly* 80 (4): 793–808.

Brechin, S., and W. Kempton. 1997. Beyond Postmaterialist Values: National Versus Individual Explanations of Global Environmentalism. *Social Science Quarterly* 78 (1): 16–20.

Bromley, P., J.W. Meyer, and F.O. Ramirez. 2011. The Worldwide Spread of Environmental Discourse in History, Civics, and Social Studies Textbooks. *Comparative Education Review* 55: 517–545.

Bromley, P., and W. Powell. 2012. From Smoke and Mirrors to Walking the Talk: Decoupling in the Contemporary World. *The Academy of Management Annals* 6: 483–530.

Brulle, R.J. 2000. *Agency, Democracy, and Nature. The US Environmental Movement from a Critical Theory Perspective*. Cambridge: MIT Press.

Brysk, A. 2013. Human Rights Movements. In *The Wiley-Blackwell Encyclopedia of Social and Political Movements* (Online).

Buechler, S.M. 2000. *Social Movements in Advanced Capitalism. The Political Economy and Cultural Construction of Social Activism*. New York und Oxford.

Burke, R. 2008. From Individual Rights to National Development: The First UN International Conference on Human Rights, Tehran, 1968. *Journal of World History* 19 (3), *New Histories of the United Nations* (Sep 2008), 275–296.

Buttel, F.H. 2000. World Society, the Nation-State, and Environmental Protection. *American Sociological Review* 65 (1): 117–121.

Cairo Declaration. 1990. http://hrlibrary.umn.edu/instree/cairodeclaration.html. Accessed 21 Apr 2017.

Canan, P., S. Andersen, N. Reichman, and B. Gareau. 2015. Introduction to the Special Issue on Ozone Layer Protection: The Extraordinary Experience of Building the Montreal Protocol, Lessons Learned, and Hopes for Future Climate Change Efforts. *Journal of Environmental Studies and Sciences* 5 (2): 111–121.

Coase, R. 1937. The Nature of the Firm. *Economica* 4: 386–405.

Cobb, N. 2014. Rethinking the 'World Polity' Perspective on Global Sodomy Law Reform. In *Handbook of LGBT Communities, Crime, and Justice*, ed. D. Peterson and V. Panfil, 283–310. New York: Springer.

Cohen, J.E. 1995. Presidential Rhetoric and the Public Agenda. *American Journal of Political Science* 39 (1): 87–107.

Cole, W.M. 2012a. Institutionalizing Shame: The Effect of Human Rights Committee Rulings on Abuse, 1981–2007. *Social Science Research* 41: 539–554.

———. 2012b. Government Respect for Gendered Rights: The Effects of the Convention on the Elimination of Discrimination Against Women on Women's Rights Outcome, 1981–2004. *International Studies Quarterly*: 1–17.

———. 2017. World Polity or World Society? Delineating the Statist and Societal Dimension of the Global Institutional System. *International Sociology* 32 (1): 86–104.

Commons, J.R. 1934. *Institutional Economics*. New York: Macmillan.

Cotgrove, S., and A. Duff. 1980. Environmentalism, Middle-Class Radicalism and Politics. *The Sociological Review* 28: 333–351.

Crotty, J. 2009. Making a Difference? NGOs and Civil Society Development in Russia. *Europe-Asia Studies* 61 (1): 85–108.

Dalton, R.J. 1994. *The Green Rainbow: Environmental Interest Groups in Western Europe*. New Haven, CT: Yale University Press.

———. 2008. *Citizen Politics: Public Opinion and Political Parties in Advanced Industrial Democracies*. Washington, DC: CQ Press.

Dalton, R.J., and R. Rohrschneider, eds. 2015. *Environmental Concerns During a Time of Distress. Environmental Politics, Special Issue*, 24/4.

Deeming, C., and K. Jones. 2015. Investigating the Macro Determinants of Self-Rated Health and Well-Being Using the European Social Survey: Methodological Innovations Across Nations and Time. *International Journal of Sociology* 45 (4): 256–285.

Della Porta, D., and S. Tarrow, eds. 2005. *Transnational Protest and Global Activism*. New York: Rowman and Littlefield.

Derbyshire, D., and I.D. Derbyshire. 2000. *Encyclopedia of World Political Systems*. Vol. 1 and 2. Armonk, NY: Sharpe Reference.

Diekmann, A., and P. Preisendoerfer. 1998. Environmental Behavior. Discrepancies Between Aspirations and Reality. *Rationality and Society* 10: 79–102.

Dietz, T., A. Fitzgerald, and R. Shwom. 2005. Environmental Values. *Annual Review of Environmental Resources* 30: 335–372.

DiMaggio, P.J. 1988. Interest and Agency in Institutional Theory. *Institutional Patterns and Organizations: Culture and Environment* 1: 3–22.

DiMaggio, P.J., J. Evans, and Bethany Bryson. 1996. Have American's Social Attitudes Become More Polarized? *American Journal of Sociology* 102 (3): 690–755.

DiMaggio, P.J., and W.W. Powell. 1983. The Iron Cage Revisited: Collective Rationality and Institutional Isomorphism in Organizational Fields. *American Sociological Review* 48 (2): 147–160.

———. 1991. Introduction. In *The New Institutionalism in Organizational Analysis*, ed. W.W. Powell and P.J. DiMaggio, 1–38. Chicago, IL: University of Chicago Press.

Doebler, S. 2015. Relationships Between Religion and Two Forms of Homonegativity in Europe – A Multilevel Analysis of Effects of Believing, Belonging and Religious Practice. *PLoS One* 10 (8): e0133538.

Dorf, M.C., and S. Tarrow. 2014. Strange Bedfellows: How an Anticipatory Countermovement Brought Same-Sex Marriage into the Public Arena. *Law and Social Inquiry* 39 (2): 449–473.

Dreher, A. 2006. Does Globalization Affect Growth? Evidence from a New Index of Globalization. *Applied Economics* 38 (10): 1091–1110.

Drori, G.S., J.W. Meyer, F.O. Ramirez, and E. Schofer, eds. 2003. *Science in the Modern Polity. Institutionalization and Globalization*. Stanford, CA: Stanford University Press.

Druckman, J.N., and J.W. Holmes. 2004. Does Presidential Rhetoric Matter? Priming and Presidential Approval. *Presidential Studies Quarterly* 34: 755–778.

Dunlap, R., and W. Catton. 1979. Environmental Sociology. *Annual Review of Sociology* 5: 243–273.

Dunlap, R.E., and R.E. Jones. 2002. Environmental Concern: Conceptual and Measurement Issues. In *Handbook of Environmental Sociology*, ed. R.E. Dunlap and W. Michelson, 482–524. Westport, CT: Greenwood Press.

Dunlap, R.E., and A.G. Mertig. 1991. The Evolution of the US Environmental Movement from 1970 to 1990: An Overview. *Society and Natural Resources* 4 (3): 209–218.

———. 1997. Global Environmental Concern: An Anomaly for Postmaterialism. *Social Science Quarterly* 78 (1): 24–29.

Edwards, B., and J.D. McCarthy. 2004. Resources and Social Movement Mobilization. In *The Blackwell Companion to Social Movements*, ed. D.A. Snow, S.A. Soule, and H. Kriesi, 116–152. Oxford: Blackwell.

EEA. 2017. http://www.eea.europa.eu/soer-2015/countries/czech-republic. Accessed 28 Apr 2017.

Eisenstadt, S.N. 2000. Multiple Modernities. *Daedalus* 129 (1): 1–29.

Esping-Andersen, G. 1990. *The Three Worlds of Welfare Capitalism*. Cambridge; Princeton, NJ: Polity Press; Princeton University Press.

Evans, J.H., B. Bryson, and P.J. DiMaggio. 2001. Opinion Polarization: Important Contributions, Necessary Limitations. *American Journal of Sociology* 106 (4): 944–959.

Fagan, A. 2004. *Environment and Democracy in the Czech Republic. The Environmental Movement in the Transition Process*. Northampton, MA: Edward Elgar.

Fairbrother, M. 2014. Two Multilevel Modeling Techniques for Analyzing Comparative Longitudinal Survey Datasets. *Political Science Research and Methods* 2: 119–140.

Finnemore, M. 1996. Constructing Norms of Humanitarian Intervention. In *The Culture of National Security: Norms and Identity in World Politics*, ed. P. Katzenstein, 153–178. New York: Columbia University Press.

Finnemore, M., and K. Sikkink. 1997. International Norm Dynamics and Political Change. *International Organization* 52 (4): 887–917.

Fleiß, J., F. Höllinger, and H. Kuzmics. 2009. Nationalstolz zwischen Patriotismus und Nationalismus? Empirisch-methodologische Analysen und Reflexionen am Beispiel des International Social Survey Programme 2003 'National Identity'. *Berliner Journal für Soziologie* 19 (3): 409.

Fligstein, N. 2001. Social Skill and the Theory of Fields. *Sociological Theory* 19 (2): 105–125.

Florini, A.M., ed. 2000. *The Third Force: The Rise of Transnational Civil Society*. Washington, DC: Carnegie Endowment for International Peace: Brookings Institution Press.

FOEI. 2017. http://www.foei.org/about-foei/history. Accessed 17 Apr 2017.

Frank, D.J., S.A. Boutcher, and B. Camp. 2009. The Reform of Sodomy Laws from a World Society Perspective. In *Queer Mobilizations: LGBT Activists Confront the Law*, ed. S. Barclay, M. Bernstein, and A.-M. Marshall, 123–141. New York: NYU Press.

Frank, D.J., B.J. Camp, and S.A. Boutcher. 2010. Worldwide Trends in the Criminal Regulation of Sex, 1945 to 2005. *American Sociological Review* 75 (6): 867–893.

Frank, D.J., A. Hironaka, and E. Schofer. 2000a. The Nation-State and the Natural Environment over the Twentieth Century. *American Sociological Review* 65 (1): 96–116.

———. 2000b. Environmentalism as a Global Institution: Reply to Buttel. *American Sociological Review* 65 (1): 122–127.

Frank, D.J., and E.H. McEneaney. 1999. The Individualization of Society and the Liberalization of State Policies on Same-Sex Sexual Relations, 1984–1995. *Social Forces* 77 (3): 911–943.

Frank, D.J., and J.W. Meyer. 2002. The Profusion of Individual Roles and Identities in the Post-War Period. *Sociological Theory* 20: 86–105.

Frank, D., J. Meyer, and D. Miyahara. 1995. Individualist Polity and Professionalization of Psychology. *American Sociological Review* 60: 360–377.

Frank, D.J., and D.M. Moss. 2017. Cross-National and Longitudinal Variations in the Criminal Regulation of Sex, 1965 to 2005. *Social Forces* 95, 941–969.

Franzen, A., and R. Meyer. 2010. Environmental Attitudes in Cross-National Perspective: A Multilevel Analysis of the ISSP 1993 and 2000. *European Sociological Review* 26 (2): 219–234.

Franzen, A., and D. Vogl. 2013. Two Decades of Measuring Environmental Attitudes: A Comparative Analysis of 33 Countries. *Global Environmental Change* 23: 1001–1008.

Fukuyama, F. 1992. *The End of History and the Last Man*. New York: Free Press.

Gareau, B.J. 2013. *From Precaution to Profit: Contemporary Challenges to Environmental Protection in the Montreal Protocol*. New Haven, CT and London: Yale University Press.

Garfinkel, H. 1967. *Studies in Ethnomethodology*. Englewood Cliffs, NJ: Prentice Hall.

Gerhards, J., and M. Hoelscher. 2003. Kulturelle Unterschiede zwischen Mitglieds- und Beitrittsländern der EU. Das Beispiel Familien- und Gleichberechtigungsvorstellungen. *Zeitschrift für Soziologie* 32: 206–225.

Gillham, P. 2008. Participation in the Environmental Movement: Analysis of the European Union. *International Sociology* 23: 67–93.

Givens, J., and A.K. Jorgenson. 2013. Individual Environmental Concern in the World Polity: A Multilevel Analysis. *Social Science Research* 42: 418–431.

Goffman, E. 1967. *Interaction Ritual: Essays in Face to Face Behavior*. Chicago, IL: Aldine.

Golebiowska, E.A. 1999. Gender Gap on Political Tolerance. *Political Behavior* 21: 43–66.

Greenpeace. 2017. http://www.greenpeace.org/. Accessed 17 Apr 2017.

Greenwood, R., C. Oliver, R. Suddaby, and K. Sahlin-Andersson, eds. 2008. *The SAGE Handbook of Organizational Institutionalism*. London: Sage.

Grundmann, R. 2001. *Transnational Environmental Policy: Reconstructing Ozone*. London: Routledge.

Guérin, D., F. Petry, and J. Créte. 2004. Tolerance, Protest and Democratic Transition: Survey Evidence from 13 Post-Communist Countries. *European Journal of Political Research* 43: 371–395.

Guo, G., and H. Zhao. 2000. Multilevel Modeling for Binary Data. *Annual Review of Sociology* 26: 441–462.

Hadden, J., and S. Tarrow. 2007. The Global Justice Movement in the United States since Seattle. In *The Global Justice Movement*, ed. Della Porta, 210–231. Boulder, CO: Paradigm Publishers.

Hadler, M. 2012. The Influence of World Societal Forces on Social Tolerance. A Time Comparative Study of Individual Prejudices in 32 Countries. *The Sociological Quarterly* 53: 211–237.

———. 2013. Environmental Behaviors in a Transatlantic View. *International Journal of Sociology* 43 (4): 87–108.

———. 2015. Institutionalism and Neo-Institutionalism: History of the Concepts. In *International Encyclopedia of the Social and Behavioral Sciences*, ed. J.D. Wright, vol. 12, 2nd ed., 186–189. Encyclopaedia Entry.

———. 2016. Individual Action, World Society, and Environmental Change: 1993–2010. *European Journal of Political and Cultural Sociology* 3 (2–3): 341–374.

Hadler, M., and M. Haller. 2011. Global Activism and Nationally Driven Recycling: The Influence of World Society and National Contexts on Public and Private Environmental Behavior. *International Sociology* 26: 315–345.

———. 2013. A Shift from Public to Private Environmental Behavior: Findings from Hadler and Haller (2011) Revisited and Extended. *International Sociology* 28 (4): 484–489.

Hadler, M., and K. Kraemer. 2016. The Perception of Environmental Threats in a Global and European Perspective. In *Green European: Environmental Behaviour and Attitudes in Europe in a Historical and Cross-Cultural Comparative Perspective*, ed. A. Telesiene and M. Gross, 13–30. London and New York: Routledge.

Hadler, M., and J. Meyer. 2009. Tangled Paths to a World Culture. Contradictory Trends in Attitudes to Globalization. In *Charting the Globe. The International Social Survey Program, 1985–2000*, ed. M. Haller, R. Jowell, and T.W. Smith, 267–280. New York: Routledge.

Hadler, M., and J. Symons. 2017. *Divergent Trends in Homophobia and Its Reflection in Public Attitudes*. Paper Presented at the Annual Meeting of the American Sociological Association, Montreal, Canada, August.

Hadler, M., and P. Wohlkönig. 2012. Environmental Behaviours in the Czech Republic, Austria, and Germany Between 1993 and 2010. Macro-Level Trends and Individual Level Determinants Compared. *Czech Sociological Review* 48: 467–492.

Hafner-Burton, E., and K. Tsutsui. 2005. Human Rights Practices in a Globalizing World: The Paradox of Empty Promises. *American Journal of Sociology* 110: 1373–1411.

Hafner-Fink, M., M. Hadler, F. Höllinger, and S. Uhan. 2017. *Probing the ISSP Social Inequality Module. Results of a Study in Austria and Slovenia*. Paper Presented at the Annual Meeting of the ISSP, Lausanne, CH, April.

Hall, P.A., and R.C. Taylor. 1996. Political Science and the Three Institutionalisms. *Political Studies* 44: 936–957.

Haller, M. 2002. Theory and Method in the Comparative Study of Values. Critique and Alterative to Inglehart. *European Sociological Review* 18: 139–158.

Haller, M., and M. Hadler. 2008. Dispositions to Act in Favor of the Environment: Fatalism and Readiness to Make Sacrifices in a Cross-National Perspective. *Sociological Forum* 23 (2): 281–311.

Haller, M., R. Jowell, and T.W. Smith, eds. 2009. *Charting the Globe: The International Social Survey Programme, 1984–2009*. New York: Routledge.

Haller, M., and C. Troy. 2003. Umwelteinstellungen und Umweltverhalten im internationalen Vergleich. *SWS-Rundschau* 43: 81–105.

Hallett, T. 2010. The Myth Incarnate. Recoupling Processes, Turmoil, and Inhabited Institutions in an Urban Elementary School. *American Sociological Review* 75: 52–74.

Hallett, T., and M.J. Ventresca. 2006. Inhabited Institutions: Social Interactions and Organizational Forms in Gouldner's Patterns of Industrial Bureaucracy. *Theory and Society* 35 (2): 213–236.

Hanhimäki, J.M. 2015. *The United Nations: A Very Short Introduction*. Oxford: Oxford University Press.

Hardin, G. 1968. The Tragedy of the Commons. *Science, AAAS* 162 (3859): 1243–1248.

Hardy, M., and J. Reynolds. 2004. Incorporating Categorical Information into Regression Models. The Utility of Dummy Variables. In *Handbook of Data Analysis*, ed. M. Hardy and A. Bryman, 229–255. Thousand Oaks, CA: Sage.

Harper, Charles. 2012. *Environment and Society. Human Perspectives on Environmental Issues*. 5th ed. Boston, MA: Pearson.

Harrison, K. 2012. A Tale of Two Taxes: The Fate of Environmental Tax Reform in Canada. *Review of Policy Research* 29 (3): 383–407.

Hart, P.S., and E.C. Nisbet. 2012. Boomerang Effects in Science Communication: How Motivated Reasoning and Identity Cues Amplify Opinion Polarization. *Communication Research* 39 (6): 701–723.

Henry, L.A. 2002. Two Paths to a Greener Future: Environmentalism and Civil Society Development in Russia. *Demokratizatsiya* 10 (2): 184–206.

Henry, L.A., and L.M. Sundstrom. 2007. Russia and the Kyoto Protocol: Seeking an Alignment of Interests and Image. *Global Environmental Politics* 7 (4): 47–69.

Hironaka, A. 2003. Science and the Environment. In *Science in the Modern Polity. Institutionalization and Globalization*, ed. G.S. Drori, J.W. Meyer, F.O. Ramirez, and E. Schofer, 249–279. Stanford, CA: Stanford University Press.

———. 2014. *Greening the Globe. World Society and Environmental Change*. Cambridge: Cambridge University Press.

Homans, G.C. 1974. *Social Behavior. Its Elementary Forms*. Revised ed. New York: Harcourt Brace Jovanovich.

Horvath, R. 2016. The Reinvention of 'Traditional Values': Nataliya Narochnitskaya and Russia's Assault on Universal Human Rights. *Europe-Asia Studies* 68 (5): 868–892.

Hovi, J., D. Sprinz, and G. Bang. 2010. Why the United States Did Not Become a Party to the Kyoto Protocol: German, Norwegian and US Perspectives. *European Journal of International Relations* 7. doi:10.1177/1354066110380964.

Hox, J.J. 2002. *Multilevel Analysis: Techniques and Applications*. New York: Routledge Academic.

Human Rights Watch. 2017. https://www.hrw.org/history. Accessed 19 Apr 2017.

Hunter, L.M., A. Hatch, and A. Johnson. 2004. Cross-National Gender Variation in Environmental Behaviors. *Social Science Quarterly* 85: 677–694.

Iglic, H. 2010. Voluntary Associations and Tolerance: An Ambiguous Relationship. *American Behavioral Scientist* 53: 717–736.

Immergut, E.M. 2011. Institution/Institutionalism. In *International Encyclopedia of Political Science*, ed. B. Badie, D. Berg-Schlosser, and L. Morlino. London: Sage.

Inglehart, R. 1982. *Changing Values and the Rise of Environmentalism in Western Societies*. Berlin: International Institute for Environment and Society.

———. 1989. *Kultureller Umbruch. Wertwandel in der westlichen Welt.* New York: Campus Verlag.

———. 1990. *Culture Shift in Advanced Industrial Society*. Princeton, NJ: Princeton University Press.

———. 1995. Public Support for Environmental Protection: Objective Problems and Subjective Values in 43 Societies. *Political Science and Politics* 15: 57–71.

———. 1997. *Modernization and Postmodernization: Cultural Economics, and Political Change in 43 Societies*. Princeton, NJ: Princeton University Press.

———. 1999. Postmodernization Erodes Respect for Authority, but Increases Support for Democracy. In *Critical Citizens. Global Support for Democratic Governance*, ed. P. Norris. Oxford: Oxford University Press.

Inglehart, R., and W.E. Baker. 2000. Modernization, Cultural Change, and the Persistence of Traditional Values. *American Sociological Review* 65: 19–51.

Inglehart, R.F., E. Ponarin, and R.C. Inglehart. 2017. Cultural Change, Slow and Fast: The Distinctive Trajectory of Norms Governing Gender Equality and Sexual Orientation. *Social Forces* 95 (4), 1313–1340.

Inglehart, R., and C. Welzel. 2003. Political Culture and Democracy – Analyzing Cross Level Linkages. *Comparative Politics* 36: 61–79.

———. 2005. *Modernization, Cultural Change, and Democracy*. New York: Cambridge University Press.

International Lesbian, Gay, Bisexual, Trans and Intersex Association (ILGA). 2016. Aengus Carroll and Lucas Paoli Itaborahy. In *State Sponsored Homophobia 2016: A World Survey of Laws: Criminalisation, Protection and Recognition*. Geneva: ILGA.

International Service for Human Rights. 2010. *GA Third Committee Deletes 'Sexual Orientation' from Resolution on Extrajudicial Executions*, November 22. http://archive.today/4wUv. Accessed 15 Mar 2016.

———. 2011. *GA Restores Sexual Orientation into EJEs Resolution, Adopts Key Texts on Death Penalty, Iran, DPRK*, 23 Mar. http://www.ishr.ch/general-assembly/983-ga-reintroduces-sexual-orientation-reference-into-ejes-resolution-adopts-key-texts-on-death-penalty. Accessed 15 Mar 2016.

Jepperson, R.L. 1991. Institutions, Institutional Effects, and Institutionalism. In *The New Institutionalism in Organizational Analysis*, ed. W.W. Powell and P.J. DiMaggio, 143–163. Chicago, IL: University of Chicago Press.

———. 2002. Political Modernities: Disentangling Two Underlying Dimensions of Institutional Differentiation. *Sociological Theory* 20: 61–85.

Jepperson, R., and J. Meyer. 1991. The Public Order and the Construction of Formal Organizations. In *The New Institutionalism in Organizational Analysis*, ed. W.W. Powell and P.J. DiMaggio, 204–231. Chicago, IL: University of Chicago Press.

Johnson, E.W., S. Frickel, and S. 2011. Ecological Threat and the Founding of US National Environmental Movement Organizations 1962–1998. *Social Problems* 58 (3): 305–329.

Jones, M.E. 1993. Origins of the East German Environmental Movement. *German Studies Review* 16: 235–264.

Jorgenson, A., C. Dick, and J. Shandra. 2011. World Economy, World Society, and Environmental Harms in Less-Developed Countries. *Sociological Inquiries* 81: 53–87.

Jorgenson, A.K., and J.E. Givens. 2014. Economic Globalization and Environmental Concern: A Multilevel Analysis of Individuals within 37 Nations. *Environment and Behavior* 46 (7): 848–871.

Katzenstein, P.J. 2005. *A World of Regions: Asia and Europe in the American Imperium*. Ithaca, NY: Cornell University Press.

Keck, M., and K. Sikkink. 1998. *Activists Beyond Borders*. Ithaca, NY: Cornell University Press.

Kenneavy, K. 2012. Support for Homosexuals' Civil Liberties: The Influence of Familial Gender Role Attitudes Across Religious Denominations. *Social Forces* 90 (4): 1347–1375.

Kennedy, P. 2006. *The Parliament of Man. The United Nations and the Quest for World Government*. London: Allen Lane.

Kidd, Q., and A. Lee. 1997. Postmaterialist Values and the Environment. A Critique and Reappraisal. *Social Science Quarterly* 78 (1): 1–15.

Klandermans, B. 1997. *The Social Psychology of Protest*. Cambridge: Blackwell.

———. 2004. The Demand and Supply of Participation: Social-Psychological Correlates of Participation in Social Movements. In *The Blackwell Campanion to Social Movements*, ed. D.A. Snow, S.A. Soule, and H. Kriesi, 360–379. Malden, MA: Blackwell.

Knight, K., and B. Messer. 2012. Environmental Concern in Cross-National Perspective: The Effects of Affluence, Environmental Degradation, and World Society. *Social Science Quarterly* 93: 521–537.

Kollman, K. 2007. Same-Sex Unions: The Globalization of an Idea. *International Studies Quarterly* 51 (2): 329–357.

Koo, J.-W., and F. Ramirez. 2009. National Incorporation of Global Human Rights: Worldwide Expansion of National Human Rights Institutions 1966–2004. *Social Forces* 87 (3): 1321–1353.

Kriesi, H., R. Koopmans, J.W. Duyvendak, and M. Giugni. 1995. *New Social Movements in Western Europe: A Comparative Analysis*. London: UCL Press.

Kruecken, G., and G. Drori, eds. 2009. *World Society: The Writings of John W. Meyer*. Oxford: Oxford University Press.

Kymlicka, W. 2007. *Multicultural Odysseys: Navigating the New International Politics of Diversity*. New York: Oxford University Press.

Kyoto Protocol to the United Nations Framework Convention on Climate Change. 1998. https://unfccc.int/kyoto_protocol/items/2830.php. Accessed 10 Apr 2017.

Lauren, P. 2003. *The Evolution of International Human Rights*. Philadelphia: University of Pennsylvania Press.

Licata, L., and O. Klein. 2002. Does European Citizenship Breed Xenophobia? European Identification as a Predictor of Intolerance Towards Immigrants. *Journal of Community and Applied Social Psychology* 12 (5): 323–337.

Lipset, S.M. 1994 [Original 1959 and 1981]. *The Political Man. The Social Bases of Politics*. Baltimore, MD: The John Hopkins University Press.

Longhofer, W., and E. Schofer. 2010. National and Global Origins of Environmental Association. *American Sociological Review* 75: 505–533.

Longhofer, W., E. Schofer, N. Miric, and D.J. Frank. 2016. NGOs, INGOs, and Environmental Policy Reform, 1970–2010. *Social Forces* 94 (4): 1743–1768.

Manning, N. 1998. Patterns of Environmental Movements in Eastern Europe. *Environmental Politics* 7: 100–134.

Markham, W. 2008. *Environmental Organizations in Modern Germany: Hardy Survivors in the Twentieth Century and Beyond*. New York: Berghahn Press.

———. 2013. Greenpeace. In *The Wiley-Blackwell Encyclopedia of Social and Political Movements* (Online).

Marquart-Pyatt, S.T. 2012a. Contextual Influence on Environmental Concern Cross-Nationally: A Multilevel Investigation. *Social Science Research* 41: 1085–1099.

———. 2012b. *Environmental Risk Perception: A Multi-level Study*. Paper Presented at ASA Meeting, Denver, August.

Marquart-Pyatt, S., and P. Paxton. 2007. In Principle and Practice: Learning Political Tolerance in Eastern and Western Europe. *Political Behavior* 29: 89–113.

Marshall, T.H. 1950. *Citizenship and Social Class and Other Essays*. Cambridge: Cambridge University Press.

McAdam, D. 1982. *The Political Process and the Development of Black Insurgency*. Chicago, IL: University of Chicago Press.

———. 1986. Recruitment to High-Risk-Activism. *American Journal of Sociology* 92: 64–90.

McAdam, D., J. McCarthy, and M.N. Zald, eds. 1996. *Comparative Perspectives on Social Movements: Political Opportunities, Mobilizing Structures, and Cultural Framings*. New York: Cambridge University Press.

McAdam, D., S. Tarrow, and C. Tilly. 2001. *Dynamics of Contention*. Cambridge: Cambridge University Press.

McCarthy, J.D., and M.N. Zald. 1977. Resource Mobilization and Social Movements: A Partial Theory. *American Journal of Sociology* 82: 1212–1241.

McCright, A., and R. Dunlap. 2011. The Politicization of Climate Change and Polarization in the American Public's Views of Global Warming, 2001–2010. *The Sociological Quarterly* 52 (2): 155–194.

Meisler, S. 1995. *United Nations. A History*. New York: Grove Press.

Menard, C., and M.M. Shirley, eds. 2005. *Handbook of New Institutional Economics*. Dordrecht: Springer.

Mertig, A., and R. Dunlap. 2001. Environmentalism, New Social Movements, and the New Class: A Cross-National Investigation. *Rural Sociology* 66: 113–136.

Merton, R.K. 1968. *Social Theory and Social Structure. Enlarged Edition*. New York: Free Press.

Meyer, J.W. 2000. Globalization. Sources and Effects on National States and Societies. *International Sociology* 15: 233–248.

———. 2005. *Weltkultur: Wie die westlichen Prinzipien die Welt durchdringen*. Frankfurt: Suhrkamp.

———. 2007. Reflections on Institutional Theories of Organizations. In *Handbook of Organizational Institutionalism*, ed. R. Greenwood, C. Oliver, R. Suddaby, and K. Shalin-Andersson, 788–809. Thousand Oaks, CA: Sage.

———. 2009. Reflections: Institutional Theory and World Society. In *World Society: The Writings of John W. Meyer*, ed. G. Krucken and G.S. Drori, 36–63. Oxford: Oxford University Press.

———. 2010. World Society, Institutional Theories, and the Actor. *Annual Review of Sociology* 36: 1–20.

Meyer, J.W., J. Boli, G.M. Thomas, and F.O. Ramirez. 1997. World Society and the Nation-State. *American Journal of Sociology* 103: 144–181.

Meyer, J.W., and R.L. Jepperson. 2000. The 'Actors' of Modern Society: The Cultural Construction of Social Agency. *Sociological Theory* 18: 100–120.

Meyer, J.W., and B. Rowan. 1977. Institutionalized Organizations: Formal Structure as Myth and Ceremony. *American Journal of Sociology* 83 (2): 340–363.

———. 1991. Institutionalized Organizations. Formal Structure as Myth and Ceremony. In *The New Institutionalism in Organizational Analysis*, ed. W.W. Powell and P.J. DiMaggio, 41–62. Chicago, IL: University of Chicago Press.

Minkoff, D.C. 1997. The Sequencing of Social Movements. *American Sociological Review* 62 (5): 779–799.

Mondak, J.J., and M.S. Sanders. 2003. Tolerance and Intolerance, 1976–1998. *American Journal of Political Science* 47: 492–502.

Muench, R. 1986. *Die Kultur der Moderne. Ihre Grundlagen und ihre Entwicklung in England und Amerika*. Frankfurt am Main: Suhrkamp.

———. 1999. Europäische Identitätsbildung. Zwischen globaler Dynamik, nationaler und regionaler Gegenbewegung. In *Kultur, Identität, Europa: über die Schwierigkeiten und Möglichkeiten einer Konstruktion*, ed. R. Viehoff and R.T. Segers, 223–252. Frankfurt am Main: Suhrkamp.

Muñoz, P., and K.W. Steininger. 2010. Austria's CO_2 Responsibility and the Carbon Content of Its International Trade. *Ecological Economics* 69: 2003–2009.

Nelson, R., and S.G. Winter. 1982. *An Evolutionary Theory of Economic Change*. Cambridge: Harvard University Press.

North, D.C. 1991. Institutions. *The Journal of Economic Perspectives* 5: 97–112.

NYT. 2016. https://www.nytimes.com/2016/12/07/us/politics/scott-pruitt-epa-trump.html?_r=0

OHCHR. 2017a. http://www.ohchr.org/EN/ProfessionalInterest/Pages/CoreInstruments.aspx. Accessed 11 Apr 2017.

———. 2017b. http://www.ohchr.org/EN/AboutUs/Pages/ViennaWC.aspx. Accessed 11 Apr 2017.

Olive, A. 2015. *The Canadian Environment in Political Context*. Toronto: University of Toronto Press.

Olzak, S. 1994. *Dynamics of Ethnic Competition and Conflict*. Stanford, CA: Stanford University Press.

Oreg, S., and T. Katz-Gerro. 2006. Predicting Proenvironmental Behavior Cross-Nationally. Values, the Theory of Planned Behavior, and Value-Belief-Norm Theory. *Environmental Behavior* 38: 462–483.

Oxfam. 2017. https://www.oxfam.org/en/about. Accessed 27 Apr 2017.

Paris Agreement. 2015. United Nations. http://unfccc.int/files/essential_background/convention/application/pdf/english_paris_agreement.pdf

Parsons, T. 1972. *Das System moderner Gesellschaften*. Weinheim: Juventa.

Peffley, M., and R. Rohrschneider. 2003. Democratization and Political Tolerance in Seventeen Countries: A Multi-Level Model of Democratic Learning. *Political Research Quarterly* 56: 243–257.

Persell, C.H., A. Green, and L. Gurevich. 2001. Civil Society, Economic Distress and Social Tolerance. *Sociological Forum* 16: 203–230.

Pesendorfer, Dieter. 2007. *Paradigmenwechsel in der Umweltpolitik. Von den Anfängen der Umwelt- zu einer Nachhaltigkeitspolitik: Modellfall Österreich?* Wiesbaden: VS Verlag für Sozialwissenschaften.

Peters, G.P., C. Minx, C.L. Weber, and O. Edenhofer. 2010. Growth in Emission Transfers via International Trade from 1990 to 2008. *PNAS* 108 (21): 8903–8908.

Pierotti, R. 2013. Increasing Rejection of Intimate Partner Violence. Evidence of Global Cultural Diffusion. *American Sociological Review* 78 (2): 240–265.

Polity Project. 2015. Polity IV Project Data. http://www.systemicpeace.org/. Accessed 15 Mar 2016.

Polletta, F., and J.M. Jasper. 2001. Collective Identity and Social Movements. *Annual Review of Sociology* 27: 283–305.

Powell, W.W., and J.A. Colyvas. 2008. Microfoundations of Institutional Theory. In *The SAGE Handbook of Organizational Institutionalism*, ed. R. Greenwood, C. Oliver, R. Suddaby, and K. Sahlin-Andersson, 276–298. London: Sage.

Powell, W.W., and P.J. DiMaggio, eds. 1991. *The New Institutionalism in Organizational Analysis*. Chicago, IL: University of Chicago Press.

Putnam, R.D. 2000. *Bowling Alone: The Collapse and Revival of American Community*. New York: Simon & Schuster.

Putnam, R.D., and D.E. Campbell. 2010. *American Grace: How Religion Unites and Divides Us*. New York: Simon & Schuster.

Quillian, L. 1995. Prejudice as a Response to Perceived Group Threat: Population Composition and Anti-Immigrant and Racial Prejudice in Europe. *American Sociological Review* 60: 586–611.

———. 1996. Group Threat and Regional Change in Attitudes Toward African-Americans. *American Journal of Sociology* 102: 816–860.

Ramirez, F.O. 1997. The Nation-State, Citizenship, and Educational Change: Institutionalization and Globalization. In *International Handbook of Education and Development: Preparing Schools, Students, and Nations for the Twenty-First Century*, ed. W. Cummings and N. McGinn, 47–62. New York: Garland Publishing.

———. 2006. From Citizen to Person? Rethinking Education as Incorporation. In *The Impact of Comparative Educational Research on Neoinstitutional Theory*, ed. A. Wiseman and D. Baker, 367–387. Oxford: Elsevier Science.

———. 2012. The World Society Perspective: Concepts, Assumptions, and Strategies. *Comparative Education* 48 (4): 423–439.

Ramírez, F.O., D. Suárez, and J.W. Meyer. 2007. The Worldwide Rise of Human Rights Education. In *School Knowledge in Comparative and Historical Perspective Changing Curricula in Primary and Secondary Education*, ed. A. Benavot, C. Braslavsky, and N. Truong, 35–52. Dordrecht: Springer.

Rohrschneider, R. 1999. *Learning Democracy*. New York: Oxford University Press.

Rohrschneider, R., and S. Whitefield. 2004. Support for Foreign Ownership and Integration in Eastern Europe – Economic Interests, Ideological Commitments, and Democratic Context. *Comparative Political Studies* 37: 313–339.

Rootes, C. 2003. Environmental Movements. In *The Blackwell Companion to Social Movements*, ed. D.A. Snow, S.A. Soule, and H. Kriesi, 608–640. Oxford: Blackwell.

Rootes, C., and R. Brulle. 2013. Environmental Movements. In *The Wiley-Blackwell Encyclopedia of Social and Political Movements* (Online).

Sanders, D. 1996. Getting Lesbian and Gay Issues on the International Human Rights Agenda. *Human Rights Quarterly* 18 (1): 67–106.

Saunders, C. 2013. Friends of the Earth. In *The Wiley-Blackwell Encyclopedia of Social and Political Movements* (Online).

Schlueter, E., and P. Scheepers. 2010. The Relationship Between Outgroup Size and Anti-Outgroup Attitudes: A Theoretical Synthesis and Empirical Test of Group Threat- and Intergroup Contact Theory. *Social Science Research* 39: 285–295.

Schmidt, V.A. 2008. Discursive Institutionalism: The Explanatory Power of Ideas and Discourse. *Annual Review of Political Science* 11: 303–326.

Schofer, E., and M. Fourcade-Gourinchas. 2001. The Structural Contexts of Civic Engagement: Voluntary Association Membership in Comparative Perspective. *American Sociological Review* 66: 806–828.

Schofer, E., and A. Hironaka. 2005. The Effects of World Society on Environmental Protection Outcomes. *Social Forces* 84: 25–47.

Schofer, E., A. Hironaka, D. Frank, and W. Longhofer. 2012. Sociological Institutionalism and World Society. In *The New Blackwell Companion to Political Sociology*, ed. E. Amenta, K. Nash, and A. Scott, 57–68. Oxford: Wiley-Blackwell.

Schreurs, M., and E. Papadakis. 2007a. *Historical Dictionary of the Green Movement*. Lanham, MD, Toronto, and Plymouth: Scarecrow Press.

———. 2007b. *Historical Dictionary of the Green Movement*. Lanham, MD: Rowman and Littlefield.

Scott, W.R. 2014. *Institutions and Organizations: Ideas. Interests, and Identities*. 4th ed. Thousand Oaks, CA: Sage.

Seifert, F. 2002. *Gentechnik – Öffentlichkeit – Demokratie. Der österreichische Gentechnik-Konflikt im internationalen Kontext*. München: Profil Verlag.

———. 2008. Einhellige Abschottung und globalisierungskritische Bauernrevolte. Widerstand gegen die grüne Gentechnik in Österreich und Frankreich. *SWS-Rundschau* 48 (4): 485–504.

Selznick, P. 1949. *TVA and the Grass Roots: A Study in the Sociology of Formal Organization*. Berkeley: University of California Press.

Senge, K. 2007. Was ist neu am Neo-Institutionalismus. *Österreichische Zeitschrift für Soziologie* 32: 42–65.

Shandra, J., L. Esparza, and B. London. 2012. Nongovernmental Organizations, Democracy, and Deforestation: A Cross-National Analysis. *Society and Natural Resources* 25: 251–269.

Shandra, J., C. Leckband, L. McKinney, and B. London. 2009. Ecologically Unequal Exchange, World Polity, and Biodiversity Loss: A Cross-National Analysis of Threatened Mammals. *International Journal of Comparative Sociology* 50: 285–310.

Shorette, K. 2012. Outcomes of Global Environmentalism: Longitudinal and Cross-National Trends in Chemical Fertilizer and Pesticide Use. *Social Forces* 91: 299–325.

Sierra Club. 2017. http://www.sierraclub.org/about. Accessed 18 Apr 2017.

Skocpol, T. 1979. *States and Social Revolutions: A Comparative Analysis of France, Russia and China*. Cambridge: Cambridge University Press.

———. 2003. *Diminished Democracy: From Membership to Management in American Civic Life*. Norman: University of Oklahoma Press.

Smith, A. 1992. National Identity and the Idea of European Unity. *International Affairs* 68: 55–76.

Smith, J. 2007. The Transnational Network for Democratic Globalization. In *The Social Movements Reader*, ed. J. Goodwin and J.M. Jasper, 236–248. Oxford: Wiley-Blackwell.

Smith, P.B., and M.H. Bond. 1993. *Social Psychology Across Cultures. Analysis and Perspectives*. New York: Harvester Wheatsheaf.

Smith, J., and D. Wiest. 2012. *Social Movements in the World System. The Politics of Crisis and Transformation*. New York: Russel Sage Foundation.

Snijders, T., and R.J. Bosker. 1999. *Multilevel Analysis: An Introduction to Basic and Advanced Multilevel Modeling*. London: Sage.

Snow, D.A., S.A. Soule, and H. Kriesi, eds. 2004. *The Blackwell Campanion to Social Movements*. Malden, MA: Blackwell.

Sotelo, M.J. 2002. Prejudice Against Gypsies Among Spanish Adolescents. *Patterns of Prejudice* 36: 28–43.

Soysal, Y.N. 1994. *Limits of Citizenship: Migrants and Postnational Membership in Europe*. Chicago, IL: The University of Chicago Press.

Stacy, H. 2009. *Human Rights for the 21st Century*. Stanford, CA: Stanford University Press.

Stegmueller, D. 2013. How Many Countries for Multilevel Modeling? A Comparison of Frequentist and Bayesian Approaches. *American Journal of Political Science* 57: 748–761.

Stern, P.C. 2000. Toward a Coherent Theory of Environmentally Significant Behavior. *Journal of Social Issues* 56: 407–424.

Stiglitz, J.E. 2002. *Globalization and Its Discontents*. New York: W.W. Norton.

Stykow, P. 2014. Russland. In *Politische Systeme im Vergleich: Formale und informelle Institutionen im politischen Prozess*, ed. H.J. Lauth, S.303–S.333. Oldenbourg: De Gruyter.

Suárez, D.F. 2008. Rewriting Citizenship? Civic Education in Costa Rica and Argentina. *Comparative Education Review* 44: 485–503.

Sunstein, C. 2002. The Law of Group Polarization. *Journal of Political Philosophy* 10 (2): 175–195.

Symons, J., and D. Altman. 2015. International Norm Polarization: Sexuality as a Subject of Human Rights Protection. *International Theory* 7 (1): 61–95.

Taraska, G. 2015, Dec 15. *The Paris Climate Agreement* (PDF). Center for American Progress. https://cdn.americanprogress.org/wp-content/uploads/2015/12/15030725/ParisClimateAgreement.pdf. Accessed 10 Apr 2017.

Tilly, C. 1978. *From Mobilization to Revolution*. Reading, MA: Addison-Wesley.

———. 2004. *Social Movements, 1768–2004*. Boulder, CO: Paradigm Publishers.

Todosijevic, B., and Z. Enyedi. 2008. Political Tolerance in Eastern and Western Europe: Social and Psychological Roots, ISSP 31st Annual Scientific Meeting 2008, July, Paris.

Tolbert, P.S., and L.G. Zucker. 1999. The Institutionalization of Institutional Theory. In *Studying Organization. Theory and Method*, ed. S.R. Clegg and C. Hardy, 169–184. London, Thousand Oaks, CA, and New Delhi: Sage.

Tsutsui, K. 2004. Global Civil Society and Ethnic Social Movements in the Contemporary World. *Sociological Forum* 19 (1): 63–88.

———. 2006. Redressing Past Human Rights Violations: Global Dimensions of Contemporary Social Movements. *Social Forces* 85: 331–354.

Tsutsui, K., and C. Min Wotipka. 2004. Global Civil Society and the International Human Rights Movement. *Social Forces* 83: 587–620.

Tsutsui, K., and Hwa-Ji Shin. 2008. Global Norms, Local Activism and Social Movement Outcomes: Global Human Rights and Resident Koreans in Japan. *Social Problems* 55: 391–418.

UIA (Union of International Associations). Various Years. *Yearbook of International Organizations*. Munich, Germany: K.G. Saur.

UN. 2015. United Nations Statistical Databases. http://unstats.un.org/unsd/databases.htm

UN Conference on Trade and Development. 2015. http://unctad.org/

UN Durban Review. 2017. http://www.un.org/en/durbanreview2009/. Accessed 11 Apr 2017.

UN Treaty Collection. 2016. https://treaties.un.org/Pages/ViewDetails.aspx?src=TREATY&mtdsg_no=IV-2&chapter=4&clang=_en. Accessed 29 Nov 2016.

UNEP. 2014. United Nations Environmental Programme. Geodata Portal. http://geodata.grid.unep.ch/

———. 2015. The Emissions Gap Report 2015. United Nations Environmental Programme (UNEP) Nairobi.

UNFCCC. 2017. https://unfccc.int/2860.php. Accessed 11 Apr 2017.

Union of International Associations. 2017. http://www.uia.org/faq. Accessed 11 Apr 2017.

Uslaner, E.M. 2002. *The Moral Foundations of Trust*. Cambridge: Cambridge University Press.

Uslaner, E.M., and M. Brown. 2005. Inequality, Trust, and Civic Engagement. *American Politics Research* 33: 868–894.

Van der Hiejden, H.-A. 2006. Globalization, Environmental Movements, and International Political Opportunity Structures. *Organization and Environment* 19: 28–45.

Van Nijnatten, D.L. 1999. Participation and Environmental Policy in Canada and the United States: Trends over Time. *Policy Studies Journal* 27 (2): 267–287.

Vasi, B. 2007. Thinking Globally, Planning Nationally, and Acting Locally. Nested Organisational Fields and the Adoption of Environmental Practices. *Social Forces* 86: 113–137.

Veblen, T. 1899. *The Theory of the Leisure Class: An Economic Study of Institutions*. New York: Macmillan.

Velitchkova, A. 2015. World Culture, Uncoupling, Institutional Logics, and Recoupling: Practices and Self-Identification as Institutional Microfoundations of Political Violence. *Sociological Forum* 30 (3): 698–720.

Verba, S., N. Burns, and K.L. Schlozman. 1997. Knowing and Caring About Politics: Gender and Political Engagement. *The Journal of Politics* 59: 1051–1072.

Wallerstein, I. 2004. *World-Systems Analysis: An Introduction*. Durham, NC: Duke University Press.

Wapner, P.K. 1996. *Environmental Activism and World Civic Politics*. Albany: Suny Press.

———. 2007. Transnational Environmental Activism. In *The Social Movements Reader*, ed. J. Goodwin and J.M. Jasper, 226–235. Oxford: Wiley-Blackwell.

Weart, S. Dec 2011. *International Cooperation: Democracy and Policy Advice (1980s). The Discovery of Global Warming*. American Institute of Physics. Accessed 7 Apr 2017.

Weick, K.E. 1995. *Sensemaking in Organizations*. Vol. 3. Thousand Oaks, CA: Sage.

Weil, F.D. 1985. The Variable Effects of Education on Liberal Attitudes: A Comparative-Historical Analysis of Anti-Semitism Using Public Opinion Survey Data. *American Sociological Review* 50: 458–474.

Weldon, S.A. 2006. The Institutional Context of Tolerance for Ethnic Minorities: A Comparative, Multilevel Analysis of Western Europe. *American Journal of Political Science* 50: 331–349.

Williamson, O.E. 1975. *Markets and Hierarchies, Analysis and Antitrust Implications: A Study in the Economics of Internal Organization*. New York: Free Press.

World Bank. 2014 and 2015. World Bank Open Data. http://data.worldbank.org/. Accessed at Various Times.

Xiao, C., and A. McCright. 2012. A Test of the Biographical Availability Argument for Gender Differences in Environmental Behaviors. *Environment and Behavior*. Published Online Before Print, 9 Aug 2012. doi:10.1177/0013916512453991.

Yang, A.S. 1997. The Polls – Trends: Attitudes Toward Homosexuality. *Public Opinion Quarterly* 61 (3): 477–507.

Yanitsky, O.N. 2012. From Nature Protection to Politics: The Russian Environmental Movement 1960–2010. *Environmental Politics* 21 (6): 922–940.

Zaller, J. 1992. *The Nature and Origins of Mass Opinion*. Cambridge: Cambridge University Press.

Zhou, M. 2013. Public Support for International Human Rights Institutions: A Cross-National and Multilevel Analysis. *Sociological Forum* 28 (3): 525–548.

Zucker, L.G. 1977. The Role of Institutionalization in Cultural Persistence. *American Sociological Review* 42 (5): 726–743.

Index[1]

[1] Page numbers followed by "n" refer to notes.

M. Hadler, *The Influence of Global Ideas on Environmentalism and Human Rights*,
DOI 10.1057/978-1-137-57440-4

GPSR Compliance
The European Union's (EU) General Product Safety Regulation (GPSR) is a set of rules that requires consumer products to be safe and our obligations to ensure this.

If you have any concerns about our products, you can contact us on

ProductSafety@springernature.com

In case Publisher is established outside the EU, the EU authorized representative is:

Springer Nature Customer Service Center GmbH
Europaplatz 3
69115 Heidelberg, Germany

www.ingramcontent.com/pod-product-compliance
Lightning Source LLC
Chambersburg PA
CBHW060749150926
51501CB00004B/132
* 9 7 8 1 1 3 7 5 7 4 3 9 8 *